I0304104

John Abercrombie

The complete Forcing-Gardener

Or, The Practice of forcing Fruits, Flowers and Vegetables to early Maturity and

Perfection

John Abercrombie

The complete Forcing-Gardener
Or, The Practice of forcing Fruits, Flowers and Vegetables to early Maturity and Perfection

ISBN/EAN: 9783337069452

Printed in Europe, USA, Canada, Australia, Japan

Cover: Foto ©berggeist007 / pixelio.de

More available books at **www.hansebooks.com**

THE
Complete Forcing-Gardener;

OR THE

PRACTICE OF FORCING
FRUITS, FLOWERS AND VEGETABLES

TO EARLY MATURITY AND PERFECTION,

BY THE AID OF

ARTIFICIAL HEAT, in the Various Departments ufually conftructed for this Purpofe.

The whole difplayed, with every new IMPROVEMENT, by which this capital and curious BRANCH of GARDENING may be effected with Facility and Succefs.

By JOHN ABERCROMBIE,

Of TOTTENHAM-COURT, Gardener:

AUTHOR OF

MAWE'S GARDENER'S KALENDAR.

LONDON:

Printed for LOCKYER DAVIS, in Holborn.

M.DCC.LXXXI.

CONTENTS.

General Description and Utility of Forcing	Page 1
Hot-Wall Forcing-Frame	33
The Forcing-House	36
The Cherry Forcing-Frame	42
Vineries or Vine-Houses	47
Bark-Heat Forcing-Frame	84
Horizontal Forcing-Frame by Bark-Heat	98
Hot-House or Stove	124
Dung-Heat Forcing-Frame and Common Hot-Bed.	165

Lately Published.

THE

British Fruit-Gardener;

AND

ART OF PRUNING:

COMPRISING,

The most approved Methods of PLANTING and RAISING every useful FRUIT-TREE and FRUIT-BEARING-SHRUB, whether for Walls, Espaliers, Standards, Half-Standards, or Dwarfs:

Together with

The true successful Practice of PRUNING, TRAINING, GRAFTING, BUDDING, &c. so as to render them abundantly fruitful:

AND

Full Directions concerning SOILS, SITUATIONS, and EXPOSURES.

By JOHN ABERCROMBIE.

[Price 4s. Bound.]

THE GARDEN MUSHROOM;

Its NATURE and CULTIVATION.

A TREATISE,

Exhibiting full and plain DIRECTIONS for producing this desirable PLANT in Perfection and Plenty, according to the true succesful Method of the LONDON GARDENERS.

By JOHN ABERCROMBIE.

[Price 1s. 6d.]

Maxims and Moral Reflections,

BY THE

DUKE DE LA ROCHEFOUCAULT.

⁎ Mr. de Voltaire asserts, that these Maxims contributed, more than any other Work, to form the Taste of the French Nation; and that they are *known by Heart*.—" Till you come to know Mankind," says Lord Chesterfield, " nothing can bring you so well acquainted with them as Rochefoucoult's Maxims. I would advise you to look into them for some Moments, at least, every Day of your Life.' *Lord Chesterfield's Letters.*

[Price 3s.]

Speedily will be Published,

THE

Gardener's Pocket Dictionary:

IN A

SYSTEMATIC ARRANGEMENT

OF

All Trees, Shrubs, Herbs, Flowers, and Fruits; with their Uses, Propagation, and Culture, in the British Gardens and Plantations, Green-Houses and Hot-Houses,

Alphabetically digested, and divided into

I. Hardy Trees and Shrubs,
II. Herbaceous Plants,
III. Green-House Plants,
IV. Hot-House Plants.

COMPREHENDING

The General Practice of Gardening, and forming a Daily Remembrancer to Gardeners, Nursery-men, Florists, Seedsmen, and all Promoters of Horticulture. The whole agreeable to the Linnæan System, with the Latin and English Names.

By JOHN ABERCROMBIE.

THE COMPLETE
FORCING GARDENER.

General Defcription and Utility of forcing Frames, &c.

THE art of forcing choice fruits, Flowers, and efculent plants to early maturity, is one of the moft curious branches of gardening, and as yet little known in general practice. It has hitherto been flighty treated of by authors, but is here fully illuftrated in the various methods practifed in the moft eminent Britifh Gardens. It comprehends not only the true fuccefsful practice of forcing fruits, flowers,

flowers and vegetables, to the highest degree of early perfection, but also of forwarding and improving many curious sorts, which scarcely will attain maturity, at all, in this our northern climate, without the aid of artificial heat in the different forcing departments; whereby we imitate the temperature of all climates and seasons, so as to raise the plants of almost every country, in a flourishing state at any season of the year.

Hence appears the great utility of Forcing-departments, highly worthy the attention of every one possessed of a good garden; especially as some sorts of them may be erected and worked at a moderate expence, and with no great trouble.

This eminent branch of gardening, having of late years, only, been improved, to any degree of perfection, and consequently the practical knowledge of it being yet very much confined, we have attempted to explain it to the world,
agreeably

agreeably to the modern practice, with many new and valuable improvements, exhibiting the different constructions, dimensions, materials of heat, and method of working, in the most effectual, easiest and cheapest ways, both for private use, and the supply of the markets.

The business of forcing being accomplished by introducing the trees, plants, roots and seeds, into the different departments calculated for that purpose, each sort shall be explained under the following heads, viz.

Fire-heat Forcing-frames,
Bark-heat Forcing-Frames.
Hot-house or Stove departments.
Dung-heat Forcing-Frames.

Concerning the several kinds of Forcing-departments, we shall first exhibit some general directions concerning their different forms, and respective utility, under the denominations of Frames, Hot-walls, Forcing-houses, and common Hot-beds.

Those denominated Forcing-frames, Hot-walls, Forcing-houses, &c. are fixed erections or high buildings, ranging East and West, full upon the South Sun, fronted and roofed with glass-work, being in width from four to ten or fifteen feet, by twenty to fifty or an hundred feet long, or more, and from eight to twelve feet high behind, with the glasses, and sometimes ranging in one slope from the front, a foot high, to the top of the back wall, and sometimes head-high glasses erect in front, with inclined sashes ranged from thence to the back wall, constructed with the back wall Northward, and a low one a foot or two high in front, either of brick work or wood, according to the materials of heat, by which you intend it shall be worked, and the sorts of plants designed to force: For instance, if fire heat be intended, you must have brick walls, in order to admit of a furnace, and flues, or funnels, proceeding from thence, ranging the whole

length

length of the walls withinside, conducting the heat to every part, in order to warm the whole interior air. But if designed to be worked with bark-heat, or with hot dung, you may dispense with strong planking; and when intended to be worked by hot dung against the outside, boarding will admit the heat more effectually; observing, of each kind, that the front be wholly of glass sashes, as also both ends, to admit of all the sun and light possible. In the narrow departments of four, five or six feet wide, designed in the Hot-wall manner, for only one row of trees behind, the glass work may be of one continued slope from the front, a foot high, to the top of the back wall: but in wide erections, intended for Forcing-houses for trees, both behind and in front, have upright glass work, head high, and from the top of it inclined or sloping sashes, carried to the heighth of the back wall; in all which departments the glasses are made to slide

slide open, in order to admit fresh air, and also to move entirely away occasionally.

Sometimes the fixed Forcing-departments, to save expences, are erected against a South wall already built, ranging the glasses as before described, in one slope to the top of the wall, or upright in the front, and sloping sashes at top; and sometimes if the wall is low, or to save expences in glazing, a back roof of tyling, &c. is carrried up from the top of the wall, forward about one third way towards the front, meeting the top of the sloping sashes, and raised high enough in front to admit the sun freely in every part; but those with the glasses continued quite home to the back erection, which admits the perpendicular light equally in every part of the frame, are much the best and beneficial to the fruit-trees, &c. behind.

In all these departments the fruit trees are planted against the back wall, and occasionally in the front, in borders formed

of

of the richest earth, and trained generally to a slight wooden trellis; and also as standards, both dwarfs, and standards with small heads, some in pots, and sometimes placed fully in the ground, where the whole bottom space is of earth; also as half, and full standards, with tall stems elevating the head near the glasses, the heads being trained within a moderate compass, not exceeding two or three feet, having, for this purpose, young trained trees of four or five years growth arrived to a bearing state; and, if planted in November a year before, they will be better prepared for forcing, which is generally begun in *January* or *February*, either by a furnace, as in all the fire heat frames, passing along the flues within, or by the assistance only of tanner's bark heat, by a back bed made in a pit intirely the whole length of the building; or with hot horse-stable dung applied either against the outside of the back of the frame, or sometimes formed into a bed in a pit withinside;

withinside; all which will be explained according to the different constructions of the respective departments.

Flowers and plants intended for forcing, may have that business effected in forcing frames, forcing houses, &c. also in bark or dung-pits, and in common hot-beds, under moveable garden frames and glasses. By these different modes of forcing, a vast variety of the most curious and desirable sorts may be raised to the greatest perfection at an early season of the year, when they will prove acceptable rarities, either for private use or public supply.

Forcing bark-pits, are constructed differently from the fore-mentioned kinds, in the open ground; these form a deep pit, close on every side, with sliding glasses at top, from four to five or six feet wide, length at pleasure, five or six feet deep behind, by four or five in front, part sunk, and the rest above ground; formed either by a nine inch brick-wall, or strong post and planking,

planking, with glafs fafhes made to fit the top; and is filled a yard in depth with bark. Sometimes a pit is formed only a yard deep on every fide, in order to be augmented occafionally with a common Hot-bed frame placed at top, together with its proper glaffes.—Thefe forcing bark-pits are filled either wholly with Tanner's bark, a yard deep, or with hot dung and bark together, the dung firft, two or three feet thick; and when it has fettled a few days or a week, add a foot of bark at top, in which plunge your pots. In either of thefe beds, of bark, or dung and bark mixed, in *January, February* or *March*, you may plunge pots of plants, feeds or roots of curious flowers, and choice efculents; fome forts alfo, as melons and cucumbers, may be planted fully in the bed, having a little earth at top; alfo pots of fmall fruit plants, dwarf-cherries, currants, &c. pots of ftrawberries, rofes and the like.

Hot-house or *Stoves*, are generally the moſt capacious artificial heat departments of any, and prove remarkably uſeful in gardening, both as a repoſitory for the pine-apple and other curious and tender exotics of hot countries, and occaſionally forwarding alſo many ſorts of hardy plants, both fruits, flowers, and eſculients. They are commonly twelve or fourteen feet wide, ten or twelve high behind, by ſix in the front, with the walls all of brick-work, to admit of flues for fire in the winter, and a capacious pit within, in which to have a bark-bed all the year, thereby to effect a certain degree of artificial heat the year round, calculated for the culture of the Pine-apple, and all ſimilar tender natives of diſtant hot climates, that are unable to endure the full air in theſe our northern parts, but require the conſtant aid of a continual warmth, equal to that from which they were originally obtained.

The

The heat required for a Hot-house, is determined by a thermometer, having the proper degree marked thereon, and it must always be such as to raise the spirit in the tube of the thermometer to the Pine-apple mark, or within 5o° over or under, to effect which, you must have fires at night, and in very cold weather during the winter and spring, from November till April or May; and always a constant bark-bed pit both winter and summer, which imparts a growing warmth at all times; and which also serves, to plunge in the pots of the more tender kinds of exotics.

The Pine plants must be always plunged in the bark-bed, for they will not fruit kindly, nor in any degree of size or perfection, unless they are indulged with that moist warmth, which is peculiar to these beds.

Besides the main hot house or stove for Pines, &c. it is right to have one or two smaller departments for forwarding the lines young Pine plants to a fruiting state, ready for the large stove or fruiting house, viz. a narrow bark pit in the full air, covered at top with glasses, serving to nurse the young plants of the year; and a larger pit or a sort of small stove, called a succession-house, in which to keep the year-old plants till arrived to full size of two years growth, then to be placed in the largest stove in autumn, to remain to fruit the next summer.

Where there are large collections of different stove plants, it is of much utility to have exotic stoves distinct from the Pine-apple Hot-house, unless indeed the Pine-house is proportionally capacious to admit of both without confusion. Though many eminent gardens have two or three principal stove departments, a Pine stove, for the culture of Pines, an exotic stove for all tender plants in general, which is sometimes divided into two divisions, one as a dry stove without bark-bed heat, principally for

for succulent plants, or such whose stems and leaves are of a fleshy nature, full of moisture, and which succeed best in a dry heat; with another division, both for fire and bark-bed-heat, in which, to keep all the tree and shrub exotics, as well as tender herbaceous kinds, that are not succulent.

The most simple and cheapest forcing frames for small plants, some sorts of fruits, and numerous flowers and esculents, are the COMMON MOVEABLE WOODEN GARDEN FRAMES, adapted principally for defending horse-dung hot-beds; and which frames, in dimensions, are commonly ten feet and half long, by four and half wide, eighteen inches, or two feet deep behind, by nine twelve or fifteen in the front, with these sliding glass-lights at top; though there are also shorter frames of the same form, for one and for two lights, by way of nursery frames, in which to raise various seedling plants for final transplantation into larger three-light-frames. So that according

ing to the dimensions of these common moveable frames, hot-beds of dung two or three feet or more high, are to be formed for one, two or several frames in ranges, extending east and west full to the sun; being made three feet and an half high for early work, two feet and an half for the more advanced season, and not less than half a yard at any season.

The utility of these common dung hot-beds under such frames as above, is very considerable in forwarding numerous sorts of small plants, sometimes fruits, particularly early strawberries, which may be raised to perfection in April; but they are principally calculated for forcing and preserving many choice esculent plants of the kitchen garden, and many sorts of flowers. The general season of making such beds, is from about Christmas until April; early beds, in January and February, are commonly for cucumbers and melons, asparagus, sallading, early radishes; and in February and March, for kidney beans, and the choice kinds of tender annuals, small beds being made first as seed-beds, to raise such plants from seeds as require transplanting, such as cucumbers, melons, and annual flowers; others of a larger size, made in a week or fortnight after, in which to prick the plants in, if the small bed is not sufficient, and in a month after, large beds for the three light frames to receive the plants of cucumbers and melons finally; with a bed also to forward the annual plants to a large size, and protect them till the season shall admit of placing them in the full air.—But several plants want not transplanting out of the bed where first placed, as asparagus, sallading, radishes, &c. Many perennial flowers may also be potted and forwarded in bloom in these dung hot-beds.

Fire-heat Forcing-houses may be continued both to be worked with fire heat, and with assistance of a bark pit, to be filled wholly

wholly with bark, or to save expence, the greater part may be of hot horse-dung a yard deep, and when settled, covered with about a foot depth of bark, which not only communicates a kindly heat, but admits also of having pots of curious plants plunged in the bark for early maturity.

In some places an entire Fire-heat Forcing-frame, where fuel, either coal, wood, &c. is plenty, and moderately cheap, is more convenient than a Bark-heat Forcing-frame; for since Pine-stoves, in which a vast quantity of bark is used, are become so general, the bark is greatly raised in price, and sometimes becomes very expensive where there is a considerable extent of pit to fill.

If hot-walls, are designed principally for a single range of fruit-trees, along the brick wall, are only four, five, or six feet wide, and to be extended fifty or a hundred feet long, it is best to work wholly by fire-heat, not with bark-pit at all,

all, having a furnace for fire at every forty or fifty feet, each fire place having its set of flues arranged along the back wall in three or more returns. Sometimes such a frame is erected against a common wall without flues, and worked with hot dung piled up thickly against the back of the wall; it is not however so certain and effectual as fire-heat; this kind of forcing-frame, though narrow, admitting only of one row of trees behind, may yet also have low plants in pots before, and pots of strawberries and the like near the front; and some vines trained in from without, and carried up thinly along the inside of the glasses.

The copious fruit houses, formed hot-house fashion, twelve or fifteen feet wide or more, either without or with a bark-pit, admitting of a row of trees both behind and in front, with other plants in the intermediate space, are rarely extended so considerably as the Hot-wall department,

ment; forty or fifty feet is a reasonable length, which may be worked with one fire profitably, by having the advantage of a large space between the back and front rows of trees to introduce pots of dwarf fruit-trees, or to plant fully in the ground, if the whole bottom space is earth; as also strawberries, both in pots and fully in the borders, together with other low plants, flowers as well as esculents; or sometimes in capacious fruit-houses, having the whole bottom space one continued plat of earth, standard fruit-trees with five or six feet stem are planted, particularly dukecherries, ranged in rows from south to north; or if the fruit-house is accommodated with a bark-pit, it may still prove of more advantage in forwarding smaller plants, plunged in the bark-bed.

Observe, in those Forcing-frames, Hot-houses, &c. designed to be worked by fire-heat, there must be a furnace of brick-work, either at one end, or in the back wall

wall behind, in which to burn the fuel for communicating the fire-heat, by means of flues or funnels of brick-work attached to the inner walls, being ranged along the front and ends to the back wall, there carried in several returns one over the other, terminating in the vent or chimney at one end, to discharge the smoak after having heated the flues, all which every bricklayer may regulate.

In those designed to be worked by bark or dung only, no fire place or flues will be required, only a six feet wide pit, by three feet six inches deep, and in length proportionable to the department, for the reception of the bark or dung-bed.

For fruit-tree forcing, &c. the fire-heat departments, worked by fire alone, are proper; by which we often ripen our choicest wall-fruits in great perfection, at an early season, keeping the heat to a certain moderate temperature, as hereafter shewn, about 10° under Pine-apple heat, though grapes succeed abundantly well by Pine stove-heat; but a fire-heat forcing-house having also the assistance of bark or dung formed within in a pit, will require less force of fire, and the general warmth will be rather more natural and kindly, both for the vegetation of fruit-trees and all other plants that you may occasionally force in the same department.

Bark heat alone without fire, is the next in utility for forcing early fruits, or indeed, for any general forcing-house; and sometimes may probably be more eligible than entirely fire heat, for its steady kindly durable warmth, with less trouble than any other artificial heat, and may likewise in some places be attended with the least expence, where fuel is scarce and dear, and bark plentiful and easily obtained at a moderate rate: on the other hand, in some places, fuel is plentiful and cheap; and bark is an expensive material, being to be had only at a great distance, so that the carriage home in any considerable quantity comes high; in all of which circumstances every one will suit his convenience; though I would remark, that if both can be readily obtained, and that if bark or dung is to be used in a pit, fire-heat conjunctively will be very effectual, &c. A moderate fire in the night, expels the cold and damps, and we may encrease the heat more or less as the weather may require, or as the trees and plants are to be more or less forwarded in growth.

Dung-heat, as the cheapest and most easily obtained, may be successfully used in forcing both fruit-trees, and most other plants, as well as for common hot-beds. For this purpose, fresh horse stable dung may always be readily procured in sufficient quantities in most places, either of your own or by purchase. This is not in all cases so effectual as fire and tanner's bark heat; however, in default of these, it may be employed with success in almost any Forcing-department, except hot-house or stoves: but for common hot-beds in

forcing

forcing kitchen-garden plants, and annual flowers, it is of singular use, and is the most convenient material for this particular branch of gardening.

The forcing departments for fruits and flowers to early growth, till succeeded by those in natural ground, require the aid of artificial heat, from January or February, till April or May.

Pine-apple stoves worked by fire and bark, effecting a constant heat the year round, serve as repositories for tender plants; some persons, who are not accommodated with other artificial heat departments, make them serve occasionally as Forcing-houses for various plants, and some small fruits, as strawberries, and many kinds of flowers, and some choice esculent plants, as kidney-beans, cucumbers, &c. These being placed in pots, and disposed in vacant parts of the house, will succeed remarkably well in their respective growth: but as to fruit trees, the Fire-
heat

heat is rather too much for most sorts, except grapes, which arrive thus to great perfection. They should always have vines planted on the outside, to be trained under the glass-work. They may have also some four or five year old vines in large pots, to move into the house occasionally in January or February, and placed near the front, or end glasses. You may also try dwarf duke cheries, currants, peaches, &c. in pots, and placed therein about February, next the front, in the most airy part, and if they produce only a few early fruits, they will be acceptable, and if they do not succeed, by reason of the heat being too considerable, there can be no loss sustained. However, the best general Forcing-frame for fruits, flowers, flowering-shrubs and esculents, is that constructed in the Hot-house-fashion, twelve or fourteen feet wide in the clear, ten or twelve high behind, and five or six in front, furnished either with flues for fire, or a bark-

pit

pit for a hot-bed of bark or dung; and if furnished also with fire-heat, the forcing may be accelerated more at pleasure, by making the fires stronger or weaker. If the bark or dung heat in the pit be sufficient, then no fire at all: having the trees planted in a border, behind, &c. and the various smaller, in pots plunged in the bark or dung-bed, and placed upon shelves at top of the flues, &c. But when intended principally as a fruit house, or to be worked by fire heat only, having the whole bottom space of earth, it may either be narrow, five or six feet in width, any length required, with one row of trees planted next the back-wall, and smaller plants in front; or may be wide, hot-house-like, with trees both behind, front, and middle space; and smaller plants between, both fully in the ground, and in pots.

The proper sorts of fruit trees for forcing are, principally *peaches, nectarines, apricots, cherries,* and *grapes*; also a tree or
two

of early figs: likewise for variety, some *dwarf currants* and *gooseberries*, with pots of rasp-berries; and always plenty of strawberries, both in pots, and in the borders.

Trees for the purpose of forcing, may be had in the greatest perfection in most of the public nursery gardens, properly trained, both as wall-trees, to arrange against the back-wall or occasionally in the front, and as standards, both dwarfs and others, with moderate heads to plant in the middle space; all of which should be young trees, not less than three, nor more than six years old, and such as are arrived to a fruitful state, and bear in tolerable abundance, according to their age and size; generally making choice of such as are of a good regular growth, moderately strong, but no ways rank or luxuriant shooters; those of the wall-tree kind with regular spreading heads full of bearing wood, and with the heads but of moderate expansion: likewise the standard trees, both dwarfs and full half standards, should have very moderate heads, not above two or three fee spread; the taller standards, and the dwarfs, less in proportion: which, if necessary, should be accordingly reduced to order at planting, and occasionally afterwards: for they must not be permitted to spread at large in these departments.

It is of importance to have the above trees, all planted a year before you begin forcing them, that they may first be well rooted and firmly established, whether fully in the ground or planted in pots; unless they can be transplanted with balls of earth about their roots, so as not to feel their removal; in which case they might be forced the same year: they being planted in November.

As to the other sorts of plants proper for forcing, there are many kinds of eminent flowers and esculents.

Of

Of the flower kind, you have the desirable fibrous-rooted flowers, as *polyanthus, narciffis, jonquils, early tulips, hyacinths,* &c. planted in winter or spring, some in pots, others in water-glasses, placing the former upon shelves or plunging them in the bed, &c. but the glasses place principally upon shelves, or any where towards the front or end-glasses. Likewise pots of pinks, stock-gilli-flowers, wall-flowers, carnations, double-rose campion, and various fibrous-rooted perennial flowers, for ornament or fragrance. Also roses and other ornamental flowering shrubs in pots; all of which being previously potted, may be placed in any forcing department, or in the hot-house, &c. in January, February, March, and April.

Kitchen esculents may be had early, kidney beans and cucumbers, dwarf peas, beans, small sallading, young mint; and, in the common hot-beds, early asparagus, melons, radishes, lettuce, carrots;

likewise in the hot-beds you may forward young cauliflower plants, cofs-lettuce, and celery, for early transplanting into the open ground; or the same sort of plants may also be still more forwarded, if necessary, by pricking them out upon gentle hot-beds for two or three weeks.

So that those who have the advantage of any kind of forcing departments, pine-stoves, hot-bed, &c. have the opportunity of obtaining many choice varieties at early seasons, long before they can be possibly obtained in the natural ground; and of having many rare plants and fruits which by no art can be made to grow in the open air in this country, as we observed before; especially in the hot-house or such fire heat departments as are provided also with a bark-pit, in which to plunge pots of various plants in the bed, in order to forward them as much as possible; also seeds and cuttings of many kinds of tender and curious plants, to facilitate

their

their vegetation, and which probably would not grow at all without the affiftance of artificial heat.

By this aid, we can in fpite of the rigours of the moft inclement feafons, obtain varieties, both of fruits, flowers, and culinary vegetables in full maturity, in February, March, April and May, which do not in their natural ftate attain perfection till June, July, or Auguft.

By this affiftance we obtain early ftrawberries in February, March, and April; cherries and currants fo early as April and May; apricots, peaches, and nectarines, in May and June; ripe grapes early in June and July; cucumbers in February and March; melons in April, May and June.

After thefe neceffary defcriptionss, preparatory to the general bufinefs of forcing, we now proceed to exhibit the feveral different methods under their refpective heads, viz. *Fire-heat forcing-frames,—Bark forcing-frames,—Hot houfe and ftove,—And Dung-heat frames,* &c.   FIRE

FIRE HEAT FORCING-FRAMES.

BY Fire-heat departments, a vast variety of vegetable curiosities may be obtained at almost any season of the year, and the curious plants of the distant warm climates with but little trouble; as by the same heat and general culture, we can maintain not only the more tender exotics in a flourishing state, but can also raise our own more hardy kinds to perfection at an uncommon season, such as the finest wall-fruits, the greatest variety of the most elegant flowers, and any of the choicest delicacies of the kitchen garden.

A *Forcing-frame* designed to be worked with fire heat, either fire alone, or jointly with bark, or dung and bark together; may, in the former case, be a narrow erection four or five feet wide, designed as a hot-

hot-wall, for only one row of trees; or a more capacious department, hot-house-fashion, as a forcing-house, to admit of a row of trees against the back wall, and another in the front, together with many smaller trees and plants in the space between the two rows; but if designed also to be assisted with bark or dung heat, it must be a wide department to admit of a six or eight feet wide pit within, in which to make the bark-bed.

Then observing in either of these methods, that all fire-heat houses must have the back front, and end walls, all of brick-work, to admit of the furnace and flues for fire, which would be dangerous in wood; the back wall to be from six or eight, to ten or twelve feet high, and one or two in the front, and at both ends; with a furnace for the fire at one end without, or behind; and flues within, running in several returns along the back-wall.

If intended to force with fire-heat only, you may have the whole bottom space within, entirely of earth, a good loamy soil, for the reception of the trees entirely in the full ground, with also some dwarf cherries, &c. in pots placed or plunged in the earth; with other small plants set fully therein.

But when it is designed to assist with bark or hot-dung, a pit of six feet wide, by three deep, must be formed within side along the middle space, or rather towards the front, extending almost the whole length of the house, sunk as much as the nature of the soil will admit, without being wet at bottom, having a four feet border behind, for the row of trees, and if there is room, a narrow one in the front for vines; and the pit being filled with bark or dung, admits of plants in pots plunged therein, with pots of dwarf-trees to move in and out occasionally, thereby having the advantage

tage of introducing fresh trees, keeping some in pots for that purpose.

The Fire-heat Forcing-frames are of two or three different kinds; denominated *Hot-Walls, Forcing-Houses, Hot-Houses, Stoves*; each as below.

HOT-WALL FORCING-FRAMES, are narrow upright fixed erections, four, five, or six feet wide, having the whole bottom space entirely of earth, designed principally for only one row of trees against the back, which may be of any length from ten to twenty, fifty or an hundred feet long or more, built with brick, eight, ten, or twelve feet high, furnished with returns of flues for fire, with a low wall in front, one foot high, on which is laid a plat of timber, thence are ranged glass frames, in one continued slope on proper beams to the top of the back wall, there received into timber work, water tight, generally having the lights disposed in two sliding tiers, the upper one made to slide up and down

over

over the under to admit fresh air, or made to slide side-ways past one another, and both tiers made to move away entirely when the season of forcing is over, that the trees may have the benefit of the full air, rain, &c. to strengthen and prepare them for future bearing, and if thought convenient, the whole frame work of wood may be contrived to move further along to a distant part of the wall, furnished with flues as above, and the frame-work remain fixed, and with one set of glasses to save a double portion of frame and walling, the wall planted all the way with trees, in order to force one half one year, the other half another, alternately; that each portion having a year's respite from forcing, they will more effectually recover proper strength, than if successively forced one year after another.

A frame of forty or fifty feet long may be worked with one fire, but if much longer, two will be necessary.

The whole bottom space of this frame being of good rich earth, two spades deep, let it be properly digged to receive the trees. Procure those that are trained in the wall-tree manner, arrived at a good bearing state, and plant them in a single range behind, about half a foot from the back wall, to admit of a trellis, between them and the wall, on which to train the branches, left, if trained close to the wall, the heat of the flues may scorch the leaves and young fruit. These being planted in October or November, with balls of earth about the root, so as not to feel their removal, may be forced the same winter. However, if they have precisely a year's growth in the forcing frame, exposed all the time to the full air, they consequently will be the better prepared for the business of forcing.

In this frame, you may have pots of strawberries, or small dwarf fruit trees, but not to shade those behind; also vines planted on the outside, and the stem drawn in thro' a hole at bottom, and laid up againſt the inſide of the glaſſes.

This frame is to be worked wholly by fire, beginning in January or February, as directed under the general head.

A *Forcing-houſe*, by fire heat, is a capacious building, ten, twelve, or fifteen feet wide or more, as much in height in the back-wall, and full hand high in the front, which, conſiſting of a dwarf wall only two feet high, the ſame at each end, and on which is erected upright glaſs-work, three or four feet high, made to ſlide open; and with inclined or ſloping lights continued from the top of the front, to that of the back wall, diſpoſed in two tiers to ſlide up and down: a fire place being withinſide at one end, or behind the back wall; from thence proceed the flues nine or ten inches wide, extending along the inſide of the front wall in one range, thence along the ends to the back wall, where dung may be continued in two or more returns, the up-

permoſt

permoſt flue running into the chimney at one end.

Obſerve, if intended entirely as a forcing houſe, to have the whole internal bottom ſpace of earth, two good ſpades deep, and light pliable loam, or any good rich garden earth, in which to plant the trees fully in the ground at once.

Or if deſigned to have a bark pit to aſſiſt with a bark or dung hot-bed, and in which alſo to force ſtrawberries, flowers, and other ſmall plants in pots, allow a ſpace of ſix or eight feet wide for the cavity of the pit, and another ſpace for a four or five feet border along the back wall, or if a narrow bark pit only five or ſix feet wide, you may have alſo a narrow border in the front, the bark pit being continued nearly the length of the houſe a full yard deep, ſunk half in the ground, if not wet; the other half raiſed, having the borders in proportion; the wall of the pit being brick, which is

beſt,

beſt, from four to nine inches thick, or of ſtrong planking well put together.

In this kind of forcing houſe, plant peaches, nectarines, apricots, cherries and vines, and a tree or two of choice plants, having been firſt trained in the nurſery-ground, and have arrived to bearing, well furniſhed with bearing wood, planted in a row in the border againſt the back wall, and trained to a trellis; and in wide departments, which admit a border in front. Plant trees next the glaſſes, or a row of grape vines, in order to be trained up by lights: otherwiſe have ſome vines planted on the outſide to train in for that purpoſe; and if there is no bark-pit within, but a continued ſpace of earth, plant therein ſome dwarf ſtandard cherries, &c. or ſome in pots plunged in the earth; alſo pots of currants, gooſeberries, raſpberries, roſes, &c.

If there is alſo a bark-pit, it may be occupied to great advantage with pots of ſtrawberries, and of dwarf trees and ſhrubs, and any

any kind of flowers, not employing any high plants to shade the fruit-trees behind. The bark will continue its heat three months; and if at the end of that time it is forked over to the bottom, it will renew its heat; or if a little fresh tar is added, it will augment the heat in a more lively degree, and be more durable.

A trellis of light post and railing must be erected close along the front of the back wall to which the range of trees there planted are to be trained, in the manner of wall-trees; not trained close to the wall, the heat of which, by the flues, might damage them; have the trellis for this purpose, formed upright three inches thick by one broad, placed about a yard asunder, well secured to the wall with hold-fasts; and have inch thick cross bars arranged horrizontally, nine inches a-part; likewise if the department is capacious enough to admit of a row of trees or vines in front, have also a
slight

slight low trellis next the glasses in this part.

At the proper planting season, the beginning or middle of September, procure proper young trees, previously trained in a fanned manner, and not less than from three or four to seven years old, arrived to a tolerable good bearing state, which may be obtained in the public nursery grounds in great choice, from half a crown to five shillings *per* tree, ready trained, but the standards much cheaper; being careful to have them taken up with as much root as possible; or when they can be raised and conveyed with a ball of earth about their roots, it will be of much advantage, but particularly, if designed to force them the same year; however, this cannot be readily effected, unless the trees are within a small distance, or have been planted in large pots a year before, or in large baskets with the balls of earth for carrying them into the destined place; but in default of such, let
others

others be taken up with their full spread of roots, and let those designed as walltrees be planted against the trellis work, six or eight feet distant, giving a pot of water directly to settle the earth about the roots and fibres, and prepare them for soon taking fresh root; then give any nursery pruning to retrench irregular shoots, &c. and fasten the branches regularly to the trellis, six inches asunder; but double that distance for vines.

Between the back and front range of trees, if wholly a vacant space, without any bark-pit, entirely of earth, may be placed dwarf standard trees, some planted fully in the earth, others in pots, such as dwarf duke cherries, peaches, apricots, or any of the fructiferous shrubs, as currants, gooseberries, raspberries, likewise pots of strawberries near the glasses; as also pots of any flowering plants; pots of kidney beans, small sallading, and in the borders,
dwarf

dwarf early peas, &c. where there is proper room without incumbering the trees.

CHERRY *Forcing Frame.*

SOME forcing departments by fire-heat are employed principally as Cherry-houses to produce early cherries: these trees not requiring so much heat, nor of so long continuance as peaches, vines, &c. and often standards, are planted for this purpose, with tall stems elevating the heads near the top glasses, for the greater benefit of the sun; you may also have half, and dwarf standards, they being all previously trained in the nursery to a bearing state, the branches well furnished with fruit-spurs: and with quite moderate compact heads, which, if at the time of planting they expand too considerably, must be shortened and reduced within a moderate compass, not exceeding two or three feet.

In this forcing-house along with the cherries, you may have a row of peaches and nectarines behind, against the back wall, trained in the wall-tree way to a trellis in regular order; and may also plant some vines in front to run up against the inside glasses; then plant rows of standard May dukecherries, cross-ways in full standard, half standard and dwarfs; having the full standard, with five or six feet stems, the half standard, three or four, and the dwarfs not more than one or two feet stems: planting them in rows four or five feet asunder, from the back to the front, the tallest behind, and so in regular gradation to the lowest in front, or have the whole planted with full standards; there will be more room for smaller plants under them, and as their branches will be elevated nearer the glasses, they may prove of advantage, in having a greater benefit of the sun to forward and improve the fruit.

In this kind of frame you may also have

various small shrubs, and plants in pots, in front and between the rows of trees. Strawberries both in pots and in the ground; or early garden-beans, peas, kidney-beans, &c.

It is of advantage to have the trees planted a year before you begin forcing, in order that they may first have taken good root: though if the trees are planted early, you may also force them moderately the first year, not beginning before February, but especially if they have been planted with balls of earth about their roots, not to feel their removal, they will more readily succeed: those however which have been a year in pots, and then placed in the forcing house, either in the pots or turned out with the ball perfectly entire, not to disturb the roots, they will succeed abundantly well by forcing the first year.

Observe, the same trees planted fully in the ground, may succeed several years in forcing, by giving them the full air always as soon as the fruit is gathered, by taking off the top glasses, to remain fully open till near the time to begin forcing; being careful as they fail or become ill bearers, to have others ready to supply their places: those however in pots may be easily so contrived, as to force some one year, and some another, by having a double portion of dwarf trees potted for this purpose; which may be forced alternately, one half one year, the other half the next: and that by relieving one another, each half, having a year's respite from forcing, remaining all the time in the full air, taking their natural growth, they will recruit their proper vigour, and be thereby enabled to bear in better perfection the next forcing season.

Trees, annually forced, do not continue so long in a healthful state, or free growth and plenteous bearing, as those growing always in the full ground and open air, according therefore as any shew an unfruitful or weak sickly habit, others should be in readiness to plant in their room, and the earth renewed with some fresh loam and rotten dung.

Observe, in general, that the top glasses are not to be put on till the time is nearly arrived for beginning the work of forcing, not however before the middle or latter end of December, suppose you intend to begin forcing early in January; unless the weather should sooner change to severe frost, when it may be proper to defend the trees occasionally, that they may be prepared by degrees for forcing; generally beginning to make the fires in January or February, sooner or later in the month, according to the time you desire to have the fruit come to perfection.

VINERIES

3. **Vineries** or *Vine-Houses by fire-heat.*

WHERE there is accommodation of different forcing frames, it is adviseable to allot one principally for the choicest kinds of grapes, such as the *frontiniac*, *muscat* of *Jerusalem*, *royal muscadine*, *Tokay*, *Syriac*, *Hamburgh*, *raisin grape*, *St. Peter's* and other large kinds, that do not ripen till late in the season in the open air, and sometimes in unfavourable seasons, not at all in England, and consequently will be greatly forwarded and now improved in size and flavour, if assisted in forcing houses by artificial heat, which generally succeed best with a somewhat larger degree than peaches and the like kinds, as they prosper in great perfection, in a pine apple stove heat, which is too considerable in general, for other trees; and as vines, in forcing, will also require the aid of shelter and

and heat longer, in order to have the grapes in their ultimate ſtate of perfection; alſo requiring a conſiderable ſcope of room to run. It is of much importance to allot them a forcing department diſtinct, where conveniency ſuits.

However, in default of ſuch convenience, plant vines in any forcing houſe, along with other trees, allowing them proper ſcope to extend and range the ſhoots a foot at leaſt aſunder.

Plant ſome on the outſide cloſe along the front, at three or four feet aſunder, then introduced them through holes, juſt at the top of the front wall, thence to be extended up along the ſides of the main rafter, and the branches conducted regularly under the lights.

Vines to be planted either within or withoutſide, ſhould be four or five years old, trained up in pots, in which they have been two or three years, and planted with the whole pot of earth entire about their roots,

whereby

whereby not sustaining any check by removal, they may be forced the same year; when they will often bear tolerably, and will encrease every year to perfection; however, in default of potted plants, procure good ones of two or three years old in the full ground, to be planted a year or two before you begin forcing, that they may have their roots well established; and they will bear the second summer, but will produce tolerably well the third. If you are not provided with proper plants of your own, they may be procured at most of the public nursery grounds; where if you obtain potted plants of some years standing, which have established their roots firmly in the mold of the pot, they may be planted, pot and all in the ground: the pot afterwards broken and cleared away, is the most expeditious method to have immediate bearing plants, but if without pots, let them be taken up in the nursery with their full spread of roots, and carefully planted with the whole entire. In default of plants, cuttings of the last year's shoots will be found to strike root, and form bearers in three years fit for forcing: but ready raised plants are most preferable.

When vines are intended to be planted separately, in a vinery or grape-house, having a broad border along the front and the back wall, and another in front, composed of a good light loamy soil, it would be of much advantage to mix it with light dry materials, such as old brick and lime rubbish, scrappings of gravelly turnpike roads, and such like dry substances, worked towards the bottom: to warm the soil, and to improve the flavour of the grapes; if the ground incline to redundant moisture below, raise the border sufficient to keep the roots from too much wet; then plant the vines along the back wall in a row, three or four feet asunder; arranged close to a trellis fixed to the wall, and to which the vines are to be trained: planting also

also another row next the front, to be conducted up to the top glasses, where must also be a sort of very light thin trellis-work, at six inches from the lights, on which to train the vines.

In front of the borders have various small plants forwarded, such as strawberries, French or kidney beans, peas, and muzagan beans, &c. as also small trees, shrubs and plants in pots; for all the vacant spaces may be employed to advantage.

The months for beginning the work of forcing are *January* and *February*; if you are desirous to try it as forward as possible, it may be begun in December, but when trees are forced too early into bloom, and severe weather succeeds, and hardly any sun to allow of the admission of a due degree of air, they are apt to miscarry: by beginning at the times first mentioned, the trees will arrive to full blossom, at a season when we may expect moderate weather and much sun, which, together with the admission of fresh air in fine days will procure abundance of success; but if too early in blossom, with little sun, weather cold, and the glasses obliged to be kept close, the young fruit is apt to drop off in their infant growth.

A week or fortnight before you begin with fire, put on all the glasses, in order to prepare the trees gradually for the artificial heat.

Observe, in forcing houses, furnished with a bark pit, or with hot dung, having a foot of bark at top, the pit should be filled a week or fortnight before you light the fires in the flues. An entire bark-bed will retain the heat the longest, but horse-dung is considerably the cheapest, and often proves sufficiently effectual in this kind of forcing; so that if you would save the expence of having all tan, procure fresh hot horse-dung, sufficient to fill the pit, in the manner as you would make a common hot-bed; keeping the glasses close, to promote

mote the heat, which will set the sap of the trees in motion, only sliding the lights a little open in the middle of sunny days; and in about ten days or a fortnight, the dung will have sunk a foot, then fill up the pit with an equal portion of tanner's bark, and the same evening begin the fire heat.

If pots of strawberries are intended, let some be now plunged in the back of the hot bed, to come earliest, others placed upon shelves to succeed them.

Remark, that whether without or with the aid of a bark and dung hot-bed, it is adviseable to make the fires pretty strong the first night, in order to expel the damps of the house and flues, and to warm them properly at first setting off; afterwards to be made moderately, every evening about four or five o'clock, or soon after sun-set, and continued till nine or ten, which will be sufficient to keep the house warm all night; observing, that, if very cold or damp foggy weather,

weather, you make a very moderate fire in the morning; or if a severe frost, keep a gentle fire, from morning till mid-day, or longer occasionally, if extremely cold; but, in moderate weather, make fires principally in the evenings; having, then a thermometer hung up in the middle of the department, serving as a guide to enable you to regulate the proper degree of heat, which should be about 60°, or if for vines entirely, from about 65° to 70°; this is principally to be understood for the artificial heat only; the sun-heat of midday will often raise it considerably higher; sometimes to 80°, or from that to 90°, in the advanced season, which is to be regulated as you see occasion, by sliding the lights at top and front, more or less, open to reduce the thermometer, to about from 60° to 70°, though it admits of a higher degree of sun-heat, than wholly by fire, which alone should not exceed 60° for a general forcing-house, or 70° for vineries or grape-houses: giving a latitude of 5° under or over, for unforeseen alterations, both of the weather, and that of the artificial heat.

If the fire-heat should at any time be too considerable, admit air, even if there is no sun, but if no sun, and the heat is not raised 5° above the allotted degree, no air is to be admitted.

However, in all fine sunny days, after the trees are proceeding in growth, and when in blossom and fruit, be careful to admit fresh air in proportion to the raised heat on the thermometer; and as the warm season advances, and the heat of the weather encreases, and as the fruit encreases in growth, give more air in proportion: the times for giving air, being from nine or ten in the morning, till three, four, or five in the evening, in which you will be governed by the natural heat of the day, or temperament of the outward air, opening and shutting sooner or later accordingly.

With respect to water, it is an article of great consequence, as rain cannot be admitted; and is necessary, both to the borders, in moderation, as you see occasion, from the earth growing dry, and occasionally all over the branches of the trees before they expand their blossom, after which, not till the fruit is set, only water the earth wherein the trees are growing moderately, once or twice a week out of the rose of the watering pot; as also about the stem of the tree; for if watered too freely all over the blossom, it might destroy the impregnating male *pollen*, or fine powder of the *anthera*, designed for effecting the office of fecundating the female organs to render them prolific: but when the blossom is decayed, and the fruit set, water all over the branches, leaves, and fruit, once or twice a week in sunny mornings, from eight or nine, to ten or eleven o'clock; but more seldom

seldom the vines, for too much moisture is hurtful to the growth of the fruit, retards the ripening and reduces the flavour; let them however have moderate supplies in the borders; and in the other trees, refresh the borders, pots, &c. those in the pots will require more frequent waterings, than those growing fully in the ground.

You will likewise observe, that pots of strawberries either plunged in the bark-bed or placed any where else, should have moderate waterings two or three times a week, according to the heat of the house, or power of the sun, taking particular care not to water too heavily upon the plants when in flower; but after the fruit is set, water freely all over the plants during their growth, but sparingly when full grown, and beginning to ripen; as much humidity will debase the flavour.

Be careful to water all other plants, either growing in the earth of the borders, &c. or in pots, those in small pots will require it two or three times a week in sunny weather, those in larger pots, and in the full borders, not so often; but give water in general, when the earth appears dry and wants moisture.

For as neither trees nor plants can receive the benefit of falling showers or refreshing dews, all necessary supplies of moisture must be effected by hand; and which must be observed with the greatest attention; heat and moisture, the great agents in this business of forcing, must be duly proportioned together.

The strawberry plants proper for forcing should be one year old at least, but if two, and arrived to a full fruiting state, it will be the greater advantage; they will bear more abundantly and larger fruit.

Take them up with balls of earth about their roots, from the best strawberry beds in the full ground, and they will bear plenteously in two months. In default of already raised fruiting plants, raise young plants

plants a year before, either from off-sets and suckers of one summer's growth from the sides of some healthful plants, taking them off towards autumn, planted six inches asunder; or take young runner plants in summer, formed from the strings or runners rooting at the joints, shut up and form young plants the same season, chusing the earliest and largest near the old stool, planting them either in a shady nursery border, or in small pots, in strong loamy soil, or any good strong rich garden mold; give the whole, as soon as planted, a good watering, and place the potted plants in a shady border all summer, being careful during their growth to clear them constantly from the summer's runners or strings, to strengthen the main head, that they may bottom well, and let them have plenty of water all summer in dry weather, more especially those in the pots.

In autumn, shift them with balls into pots a size larger; likewise those in the beds take up alfo with balls and pots, giving each fome water to fettle the earth firmly about the ball; and if ftrong well grown plants, place them in the forcing-houfe at the proper time; though they rarely bear in any abundance till the fecond year: So if not ftrong plants the firft feafon, let them have another fummer's growth managing them as before.

But to have fruiting, or immediate bearing plants fit to force directly, have recourfe in October or November, to any good beds of ftrawberries in the natural ground, that are about two or three years old, not more, juft in perfection, take up a quantity neatly, with balls of earth to their roots, and put them in middling pots (24s.) in a good rich loam, or in default of this, the beft kitchen garden mould, and give water directly; then place the pots in a fheltered funny fituation till the forcing time. Sometimes fuch plants being taken up only juft at the forcing feafon, season, if the weather is open will succeed; but it is better to have them potted a month or two before. And it should be observed in general of strawberry plants intended for forcing-heat, if severe frosty weather happens before the time of forcing them, give the shelter of a garden-frame; for it is of importance to have the plants in full vigour, not impaired by inclemency of weather.

It should likewise be remembered, that in order to have a longer succession of forced strawberries continued till the season of the natural ground crops, always have a reserve of pots of fruiting plants, some advancing in a very moderate dung hotbed under frames, others placed under frames and lights without heat; so removing them by degrees into the fruitery houses, at about three weeks or a month's distance of time, and they will then succeed one another in bearing, from February

bruary or March, until the season for crops in the full ground.

It should likewise be remarked, that it is proper to have a fresh stock of plants for each year's forcing, and never to use them if more than two or three years old; and always procure them from such beds as are noted for being good sorts, and in the most fruitful state, by no means from old beds in which you will certainly be deceived.

The best strawberries for forcing are the scarlets; the alpine or monthly strawberry and hautboys may also be tried for variety.

Pots of gooseberries, currants, raspberries and the like, placed any where in the forcing frame in January and February will succeed; the gooseberries and currants green for tarts, and some of all three sorts ripe in April or May

Roses of all sorts, *syringas*, *hypericums*, *Persian lilies*, *honey-suckles*, *oranges*, or any dwarf flowering

flowering shrubs approved of, being potted Roses the year before, may be placed in the forcing-houses any time from January to February and March, both in those without, and with, bark pits: they will blow freely early in the spring; but if desirous of having some roses, &c. as early as possible; and if there is the advantage of a bark-pit, plunge some pots in the bark, which will forward them in a pleasing manner to a fine early bloom.

Pots of *pinks, carnations, sweet-williams, tulips, hyacinths, narcissus, jonquils, amaryllis, lilies, pancratiums, anemones, ranunculus,* and any other choice kinds of harbaceous flowery perennials, both fibrous, bulbous, and tuberous rooted kinds that may be required to flower early, may be introduced in the forcing-houses in winter or early in the spring; some plunged in the bark-pit, others placed upon shelves.

Any bulbous flower-roots set in
blowing

blowing glasses of water, being placed upon shelves in the forcing-houses, will also blow in fine perfection in winter, and till those in the natural ground come into flower.

Likewise any choice annual flowers; such as *ten-week stocks, balsamines, gemphrenas, mignionette*. and other early flowers, with a showy bloom, or remarkable for fragrance; or any of the common hardy sorts which you desire shall flower early; such as pots of bark spurs, sweat peas, and candy-tuft, &c. which sow thick, to remain where sowed.

CUCUMBERS, may likewise be raised by fire-heat, in hot-houses accommodated with a bark-bed within, particularly in pine-stoves: having some young plants previously raised from seed in small pots, either in the bark-bed or in a dung hot-bed under common frames, sowing the seed in January or February, plunging the pots in the bark or dung bed, they will

will come in two or three days, and when only two or three days old, prick them out in small pots, three or four in each, laid on up to their seed leaves, the pots then plunged in the bark or hot-bed, and occasionally very moderately watered, in a sunny forenoon. When the plants are a little advanced in growth, having the rough leaf two or three inches broad, transplant them where they are finally to remain, either in large pots (16 or 24s.) or oblong narrow boxes two or three feet long by eight or ten inches width and depth, filled with rich light earth, and placed near the upper glasses upon narrow shelves towards the back part, a foot or two distance from the light; or the boxes may be of a more continued length, and suspended from the rafters of the lights by iron or wooden brackets without shelves; for it is of much consequence to have these plants near the glasses to receive all possible benefit of the sun; not placed
immediately

immediately under open sliding lights, but a little before or behind that part.

As to the future culture, give moderate watering, as you shall see occasion in warm sunny days in the forenoon; and, according as the plants shoot out runners in length, place up rods or laths for their support. Likewise, when the plants are in blossom take care to impregnate the female flower with the *anthera* of the male, which you will pluck for that purpose the same day that both flowers first expand; pulling away the flower, leaf or petal of the male blossom, then holding the shank betwixt the finger and thumb, introduce the anthera into the center of the female flower, touching the stigmata thereof, and twirl it about so as to leave some of the male powder upon the female organ, then throw it away, this completes the business, renders the flowers fertile, and the fruit sets freely. And this is called setting, which should be constantly attended to in

all

all the early fruit, taking the opportunity of effecting it the fame day the flowers, both male and female open, being then in perfection to enfure fuccefs.

As to kidney beans, peas, mazagan beans, and any other choice efculents intended for forcing, they may be cultivated in pots or in borders. They may be planted in oblong narrow wooden trough-like boxes, of a yard long, by eight or nine inches wide, and ten deep. You may either plant the beans at once, in large pots (24s.) or in the boxes finally to remain, three beans in each pot, or in a double row along the boxes, an inch deep, or previoufly in fmall pots (32 or 48s. for tranfplanting into the larger pots or boxes, with the ball of earth about their roots.

The feafon for planting them is January, February and March, to fucceed each other, full to the fun, giving but very little water when the beans have fprouted.

The

The transplantation into larger pots, should be effected when they are advanced two or three inches, being in the first rough an inch or two broad, turning them out with the lump of earth into the larger pots at nine inches distance; boxes are very convenient to arrange along the top of the parapet-wall of the bark-pit of forcing-houses and pine-stoves.

Observing, as to the culture of kidney beans forced in these departments, it is principally to supply them with water, being careful to give but very little or none till the beans have germinated or sprouted, for before that, much humidity would rot the bean, but after they have shot, and in their future growth, supply them duly with water; two or three times a week to those in pots, and those in the ground in proportion.

They will bear in six or eight weeks after planting, observing to gather them twice a week, suffering none to grow
large

large, and thus they will continue bearing longer and more abundantly.

Keep young plants advancing in small pots, sowed at three weeks intervals, ready to turn out into larger pots to succeed the old plants as they decline bearing: thereby continue a succession of beans from March or April until June, when the natural crops come in.

The sorts commonly used for this purpose are the early dwarfs, either the white, black, or liver-coloured; but the speckled dwarfs are the strongest, more plenteous bearers, and continue longest in bearing.

Dwarf peas and mazagan beans may also be raised in the fruit houses, where there is broad borders of earth, to tolerable good perfection to gather in April, they may either be sowed at once in the borders of the forcing-house in December or January, in drills an inch deep, and half a yard asunder, ranging south and north;

north; you may have them coming forward under some common garden frame, defended from frost with the glasses, or in some gentle dung hot-bed, sheltered in the same manner, being sowed quite thick, either in the earth or in pots, for transplanting; performing the sowing a month or six weeks before you intend to work the forcing frame, though if in a hot bed, if sowed only a fortnight or three weeks before hand, it is time enough then. When the forcing is begun, move those in pots into the house, which, as well as those not in pots, when an inch or two high, not more, are to be transplanted with all their roots into the borders, in rows, the pea-plants two inches apart, the beans three or four, giving directly a moderate watering to settle the earth, and to forward their rooting and free growth.

Let them, after this, have moderate refreshments of water in sunny days, once or twice a week or oftener as the warm season

season advances, and the plants encrease in growth; but when in bloom, supply them well with water in warm weather, as also when the pods are set; observing likewise when they are in full bloom to pinch or cut off the top of each plant, that all the nourishment may be thrown to the supply of the pods, in which they will set sooner, more abundantly, and attain perfection earlier.

Likewise in any of these departments, when hard winters have destroyed the natural early crops of peas and beans, you may have some forwarded to supply their places, being sowed thick, in large pots, boxes or baskets, placed near the heat; they will soon come up, when they must be gradually moved to the open air, by placing them out under a frame exposed to the full air all day, and by degrees, hardened to it night and day, then planted out close under some warm south wall, in a single row close along the bottom.

SMALL

SMALL SALLADING of any fort, may also be raised early in thofe kinds of fruit-houfes, fuch as creffes, muftard, rape, raddifh, &c. fowed thick, in large pots or boxes, to be cut for ufe while quite young, in the feed leaf, not more than a week old. You should fow fome every week or fortnight, very thick, and only juft covered with fine earth; and if placed any where in the houfe, giving moderate fprinklings of water, they will foon come up, and be fit for ufe in a few days after.

You may likewife have pots with fpeermint roots, thickly planted, an inch deep, and placed any where in the forcing-houfe, which will furnifh early mint for fallads and mint fauce.

By the affiftance alfo of forcing-houfes furnifhed with a bark-pit, you may forward the ftriking or cuttings of any curious exotics, as *myrtles*, *geraniums*, &c. planted in pots thickly, and plunged in the bark-bed.

In

In regard to the trees, they are sometimes in their early growth apt to be attacked with a fort of infectious insect, causing a pernicious blight, which if permitted to spread, will do much mischief. In this case, it is adviseable to fumigate if with burnt tobacco, in fumigating bellows*; which often proves efficacious in exterminating the vermin, but more effectually if, with the assistance of a smoking pot filled likewise with tobacco, lighted and blowed with bellows, to effect a stronger fumigation.

However, in trees slightly infected, use only the common fumigating bellows, apply the pipe towards the blighted parts, at a sufficient distance, not to scorch the leaves, &c. continuing to throw a stream of smoak against the trees for some time.

If you apprehend the infection spreading considerably, it is of consequence to fumi-

---

* A late invention; they are sold by the Braziers and Tin-men.

gate the house throughout, causing a thick smother, so as to darken the house, keeping the glasses close shut, which is of great service in effectually exterminating the infectious blighty race, proves very conducive to the health and prosperity of the trees, facilitating a kindly growth, the fruit swelling freely to its proper size, which otherwise is apt to get shorter, and drop before it attains maturity.

In neglect of the above operation, you will notice such shoots as are attacked with this blight or insect: causing the leaves and shoots to shrivel; pick off the most infected leaves, and cut down the shoots below the infected part, to prevent the disease from spreading.

Observe, if the fruit of appricots, peaches, and nectarines, should any where set in thick clusters, let them be thinned, removing the worst, and retain the most promising and fairest, sufficiently distant to give them room to swell, without

thrusting

thrusting one another off as they increase in bulk.

If any bunches of grapes have the berries considerably crouded, let the worst and ill grown, be thinned out early, with narrow pointed scissars, to give the remainder full scope of growth, as well as to admit the sun freely to ripen the whole equally.

When the forcing season is past, that is to say, when the fruits are fully ripened, and for the most part gathered, do not omit removing off all the glasses, in order to admit the full air and showers of rain to the trees, to ripen and harden the young wood more effectually, and prepare the general bearers for next winter's forcing, leaving the whole fully open, night and day, till within a fortnight of the next forcing time, when they should be put on again.

Relative to PRUNING, let the trees have this operation annually performed; especially

cially thofe which are trained wall-tree fafhion, fuch as peaches, nectarines, apricots, vines, cherries, &c. that are arranged againſt the trellis, and the vines trained up under the glaſſes; all of which will require both a ſummer and winter prunning.

In the ſummer regulation, remove the ill-placed and imperfect young wood of the year, and any fuperabundancy, being careful to train in a plenteous fupply of all the well-placed perfect ſhoots cloſe to the trellis; in peaches and nectarines, appricots, vines, and figs, which, bearing chiefly on the yearling ſhoots, and but one year on the fame wood, a general fupply muſt be retained in every part of each ſummer's ſhoots, to chufe out of in winter pruning for next year's bearers; as they all produce fruit principally on the young ſhoots, extend them moſtly at full length all ſummer, and continue them cloſe to the trellis during their growth, and until all the fruit is ripe; but cheries

and

and plums, &c. in which the same branches continue bearing several years, not wanting a general renewal of bearers annually, as peaches, &c. need at this time only, here and there, a good shoot in the most vacant spaces, trained in.

In summer-dressing the vine, observe well the bearing shoots, which, being the same year's wood, let plenty of the fruitful ones be trained in with great regularity, both to furnish the same year's grapes, and as succession wood for next year's mother bearers; divesting them of any side shoots. Then extend them in length close to the trellis; not cross one another, but all parallel in the neatest order: and when the branches of fruit are advanced in growth, to distinguish their goodness, if they are too abundant in any shoot, displace the worst and most irregular, leaving the largest and finest branches, three or four in a shoot.

In the winter-pruning you will observe of those trees ranged against the trellis, wall-tree fashion, to continue the branches regularly trained, five or six or eight inches asunder, forming a regular display of bearers, covering the spaces allotted for each tree; at the same time, to cut out worn out or blighted wood, and renew the place with young of last summer; being also careful to continue each tree within its limited bounds, so as not to interfere with another.

Vines. The time for performing this pruning, may be as soon as the leaf falls in November, especially in vines, but as to peaches, nectarines, &c. they may be deferred till near the time for beginning to force, when the fruit or blossom buds will be more advanced to enable you to make a proper choice of the best shoots.

Trees bearing on the young wood, as peaches, nectarines, apricots, &c. must retain a general supply of last summer's shoots

shoots in every part, for next year's bearers; and part of the old ones at the same time cut out in proportion, to give proper space to train the successional supply; cutting out the superabundant young shoots, if any, and shortening the reserved ones as the different trees require, particularly peaches, nectarines, and apricots; to encourage their furnishing more certainly a supply of collaterals from the lower eyes the preceeding summer, as new succession bearers, which otherwise would be apt to rise mostly towards the upper part, and leave the bottom; so that it is right to shorten each shoot in the winter-pruning according to its strength, by pruning off about one half, third or fourth of each, generally cutting each to a leaf or branch bud to furnish a leader to the bearer, or to a twin blossom, having a wood bud issuing from between, to afford also a leading shoot at the end to draw nourishment more effectually to supply the fruit. And as to vines, shorten them from three or four, to five or six joints or more, as mentioned more particularly below. Cherries and plumbs, which bear several years in the same wood upon spurs, must not have the shoots or branches shortened, unless they shoot out of bounds,

Let the whole, as soon as prunned, be regularly trained to the trellis in order, at equal distances, not less than six inches, but vines double that distance.

Standard cherries, or other fruit trees in the forcing houses, need but very moderate pruning; remembering that in those departments the heads must always be kept within a small compass, and the branches thin and at regular distances.

Every year either before or after performing the necessary winter pruning, it is proper to dig the borders; and once in two years, add some good rotten dung, or some fresh loamy compost, the whole neatly digged

ged in a spade deep, taking particular care not to disturb the roots of trees.

The dwarf trees in pots should occasionally have the earth stirred at top, and sometimes a little of the old removed and replaced with fresh loam, &c. or occasionally shifted in autumn into fresh earth, removing them with the whole mass of earth about the roots entire, trimming off little of the old earth, then having fresh compost of loam or other good rich earth in the bottom of the new pot, place the tree with its ball therein, and fill up all around, and at top add more fresh mould, and finish with a good watering.

All pots of trees, shrubs, and herbaceous perennial flowers, &c. when they are past fruiting and flowering, &c. should be moved into the full air in a shady border for the remainder of summer and autumn, loosing the earth at top, or shifting those that require it into fresh earth or larger pots: those not designed to force again next winter may be turned out of the pots into the ground, supplying them with water all summer, and keeping the whole clean from weeds.

If any choice bulbous and tuberous roots have been forced, they may be taken entirely up when the flower is past and the leaves begin to decay, in order to clear off the increased off-sets: plant them in fresh earth in autumn, either in pots to force again, or in beds for a year, and force a fresh supply the year ensuing.

Any other perennial plants, shrubby, or herbaceous, will be prosperous if they have a year of respite from forcing, having others ready potted to force at the proper season.

Sometimes fruit-trees forced in the narrow departments, called Hot-Walls, are contrived to have a year's rest from forcing, by having a double portion of walling occupied with frames; one set of glasses to serve both, made to slide or move along from one part to the other, thereby affording

ing an opportunity of forcing the trees half one year, and the other half the year following, alternately; each half having a year's refpite in their natural growth, to recover a long fhare of ftrength, in which they will have a better chance of producing more eligible crops of fruit next forcing feafon.

BARK-HEAT FORCING-FRAMES.

BARK-HEAT forcing-frames are eligible where tanner's bark can be easily obtained, so as to be cheaper than fire-heat; in which case such forcing frames are well worth attention, and will prove very successful, by forming the bark into a substantial hot-bed, in a deep oblong pit or cavity, within the forcing house, called the bark-pit.

This valuable material is the detached bark of the oak tree, chopped to pieces, for tanning of leather; and which afterwards becomes an important article in gardening, making the most eligible hot-beds in the world, being much superior for its uniform, moderate, and durable

ble temperature of heat, to horſe-dung hotbeds, which are apt to heat vehemently at firſt and ſoon decline, and not liable, like thoſe beds, to injure tender plants by rank pernicious ſteam; but ſupports a ſteady and growing heat three times as long, adopted to almoſt every ſort of vegetable growth; and the beſt calculated for forcing every kind of tree, flower, or plant, as well as for the cultivation of the pine apple, and all tender exotics.

A good bark-bed will ſupport a moſt agreeable fine growing heat, three or four months without further trouble; and if at the end of that time, the bark is forked even to the bottom, looſening and well mixing the parts together, it will renew its fermentation, and recruit its declining heat, for a month or ſix weeks longer; and with the addition at the end of five or ſix months, of a portion of freſh bark, equal to one third of the whole, working up all well together, it will renew the heat so effectually, that the same bed will retain a growing warmth, for nearly the year round.

At the end of ten or twelve months, from the first making of the Bark-bed, it requires to be renewed from the tanyard; you may previously screen the old bark, removing all the small earthy stuff that passes through the screen, then filling up the pit with fresh bark, sufficiently to allow for six inches settling, working up and blending the new and old bark together; and in a fortnight, or a little more or less time, the bed will acquire a proper degree of heat, for the reception of pots of plants, observing some precaution at first plunging, when the bed is new, lest the heat prove too violent; either be careful not to place them in the bark too soon, or not to plunge the pots more than one third or half way at first.

These

These most useful hot-beds, are calculated not only for most sorts of forcing-frames, but are the grand support of hot-houses, and Pine-apple stoves, which require the joint aid of bark-beds, and fire-heat, the latter however only in winter, but must have bark beds the year round; without which the Pine apple, and some other tender exotics cannot be raised to proper maturity.

The bark is to be procured by the cart load, or by the bushel, which, in the environs of London, is commonly at the rate of a penny or three-halfpence, and by measuring the dimensions of your bark-pit, you will easily calculate the necessary quantity required; making choice of the fresh lately cast out of the tan vats; not such as has lain long and become any ways earthy; chusing also the middling sized in preference to the small, which decays, exhausts its vigour, and becomes

becomes earthy sooner than the middling and larger sized bark. Having obtained the quantity required, observe, if very wet, being newly thrown out of the tan vats, spread it abroad in the sun and full air, or dispose it in little heaps, to drain off the superabundant moisture; then either cast it together for a few days to promote the fermentation, or, if the bark-pit is ready, carry it there at once, filling up the pit some inches above the brim, to allow for settling.

Observe, that as bark is a short loose material, it cannot be formed into a bed, without being thrown into a pit to confine it together, in the form required, which pits for the bark-beds, denominated bark-pits, as formerly observed, being of an oblong form, six or eight feet wide, by three and a half deep, and in length, proportionable to that of the Forcing-house, in which they are made, are

sometimes

sometimes sunk a foot or more into the ground; the rest above, and formed by a surrounding thin brick wall, or thick board or planking; sometimes a simple bark-pit, not constructed within any building, is formed in the open air, with glasses just to fit and cover it at top, designed principally for forcing small plants in pots, or for containing young Pine-apple plants. See *Hot-house Departments*.

Sometimes bark is employed together with hot dung to save expences, first filling the bark-pit with new hot horse-dung to the brims, and in ten or twelve days when it settles a foot, fill up the pit with the bark; which together, will effect a very kindly growing heat, but not near so durable as an entire bark-bed; it however will answer the purpose of forcing some sorts of fruit trees that soon ripen, such as cherries, strawberries, and any kinds of flowers, kidney-beans, sallading, asparagus, &c.

Bark

Bark is also used jointly with horse-dung, in forming hot-beds in the open ground, to be defended with common hot-bed frames, making the bed either with layers of dung and bark alternately, or with dung below and bark at top; in the former method, begin with a layer of hot-dung a foot thick, raised up on both sides, and at each end, to form a cavity for a foot layer of bark, and to secure it from slipping down; covering this with six or eight inches of dung, raised on the sides and ends as before, then a foot of bark at top of all, secured on the sides with the raised dung as above; this being the only method by which a bark-bed can be made without a pit, and makes a very kindly hot-bed, and of much longer duration than one of all dung; sometimes a hot-bed being made almost wholly of hot dung below, two feet thick or more, is covered with a foot of bark at top, forming a rampart of dung as above described, to contain and keep it up.

In

In thefe hot-beds, place pots of young Pine apple plants, to nurfe them, particularly the off-fet fuckers of the old ftools, and crowns at top of the fruit, in Autumn, to ftrike them for propagation, and nurfe them a month or two or more, or even during the winter, lining the outfides of the bed occafionally with hot dung and litter.

There are different kinds of Bark-heat Forcing frames, agreeable to the ufes for which they are calculated, fome for fruits, others for flowers, and fometimes efculent plants; all formed with a bark pit within, from three or four to fix or eight feet wide, and a yard deep or more, in which to form the bark-bed; fome alfo with furnace and flues for fire.

Some Forcing-frames for bark-heat are formed nearly in the hot-wall manner, againft any South fence, and only fix or eight feet wide, with a three or four feet wide bark pit, continued along towards the front, of front, with a border behind, of the same width, to admit of a row of trees trained to a trellis; another fort is formed in the Pine-houfe manner, twelve feet wide or more, with a bark pit fix feet wide along the middle fpace, admitting of a border, both againſt the back-wall and the front, for a row of trees in each; or with the pit continued forward to admit of a border behind. Other forts are contrived in the manner of a deep garden-frame, but more capacious in all refpects, with the bark-pit of the whole width, and the trees planted on the outfide with the ftem and head, admitted within, to train upon a trellis ranged parallel to the top glaffes; and fome being alfo made for a large garden-frame in a pit, in which to receive pots of trees, fhrubs and flowers.

But the moſt eligible Bark-heat Forcing-frames for general work, are thoſe of the Hot-wall and Forcing-houfe kinds.

The

The Hot-wall Frame need only be six or eight feet wide, with a three or four feet wide bark-pit, arranging towards the front, admitting of the same width behind for a border, to receive a row of trees against the back-wall, which should be seven, eight, or ten feet high, either of brick, as the warmest, or a strong thick boarded fence, closely joined, which will admit of a lining of dung occasionally, or may be formed against any south wall or close pailing already built; in either case, form a wall in front one or two feet with glasses, reaching from thence to the top of the back wall, having the bark-pit formed next the front, the whole length, and the width in proportion to that of the whole frame, and three feet deep, either mostly sunk, or if in danger of wet, one half sunk, the other raised; the wall of which being either four inch brick-work, or planked up, forming the border behind, raised more or less according to the bark-pit,

plant

plant a row of trees as in the fire-heat frames, with some vines on the outside, to train in against the inside of the glasses.

Fill the bark pit with new bark from the tanners in January or February, according to the time you desire to have the fruits, flowers, &c. in perfection. When the bed is formed, keep all the glasses close, it will draw up the heat sooner, and be of a good temperature in a fortnight, in which may be introduced pots of strawberries, flowers, roses, kidney-beans, &c. all of which will grow most kindly.

THE FORCING HOUSE by *bark-heat*, may be capacious, in the pine hot house form, twelve or fourteen feet in the clear, with a six or eight feet wide bark-pit along the middle, allowing a space for a broad border behind next the back wall, and a small one in front: the back wall being built eight, ten, or twelve feet high, or any garden wall already built,

full

full to the south, with one in front only a foot or two high; on which is placed upright glasses, four or five feet, and from thence sloping sashes continued to the top of the back wall, as observed in the fire-heat forcing house, ranging in two tiers, which, as well as those in the front, are made to slide open, and move quite away occasionally; the bark-pit the width above mentioned is a yard deep, either sunk more or less as the ground admits, but not raised too high to shade the trees behind, and may be either bricked or planked on the sides, but brick work is the best.

Trees are to be planted here in a row next the bark wall, and either a trellis fixed up, or trained immediately to the wall, if no flues. You may plant a row also next the front glasses, where you may have vines between the trees, to be conducted up under the inclined or sloping lights, at top, or on the outside, with the stems drawn in through small apertures.

You

You may begin forcing this frame in January or February, making allowance of a fortnight for the bark arriving to a proper heat, after being put into the pit: so that if the pit is filled with new tan by the middle of January, it will heat and begin to force the trees about the end of that month, and continue its heat two or three months; or if at the end of two months, it declines considerably, fork it up to the bottom, which will produce a fresh fermentation, and a renewed durable heat.

In this forcing house you may likewise have strawberries forced abundantly, having different supplies of plants in pots, placed under frames in winter, defended from frost, ready to move into the forcing-house at intervals of three or four weeks, to succeed one another, plunging some in the bark-bed for the earliest, others any where in the house near the glasses; and they will furnish as fine strawberries, as in the open ground, in their natural season;

and

and by supplies of fresh pots of plants twice or thrice in the season, you may continue a succession of fruit till the natural crops come in.

Any kind of low flowering shrubs, flower-plants and roots, potted and placed here, will blow abundantly fine and early; likewise may raise any curious annual flowers to an early bloom; sow the seed in February or March, in pots, plunge into the bark-bed, and prick out the seedling-plants into other pots singly: also seeds of hardier annuals may be sowed here in pots, to be forwarded for the open ground bloom; the young seedling plants being hardened by degrees to the full air, either for pots or the full ground.

Have also pots of kidney-beans, with some planted fully in the borders under the trees; where likewise, if there is room, you may have some rows of early dwarf peas, &c.

Also occasionally in this bark-bed, you may strike cuttings of various tender exotics. Large orange and lemon tree stocks may likewise be set in the spring.

As to the general culture or management of this forcing department, the principal care is occasional waterings, and the admission of fresh air.

A Horizontal Fruit Forcing-Frame, *by Bark-heat.*

THIS kind of frame is detached from any wall or upright erection, is constructed of wood, somewhat in the manner of a common hot-bed frame, but considerably wider and deeper, having sliding lights at top, and is placed upon a deep pit for a bark-bed, and the trees planted on the outside of the front, and trained with the branches expanded just under the glasses, in a somewhat horizontal position; it is made wholly of wood, close on every side, having a yard

yard deep bark-pit funk underneath to fit the frame; the trees being planted in a border clofe on the outfide of the front; and the ftem and head of the tree introduced at bottom of the frame, and trained to a trellis parallel to the glaffes.

This kind of forcing frame being made with inch and half, or two inch deal board, fix feet wide by ten long, three feet fix inches deep behind, by half a yard or two feet in the front, both ends in proportion; has two glafs fafhes to fhove up and down and move away occafionally. Two, three, or more of fuch frames may be conftructed to range in a line, with a bark-pit proportionally long; each frame having a fliding door behind for entrance to receive the bark, and to go in to fork it up, to renew the heat, &c.

To fit thefe frames, a bark-pit of wood or brick work is conftructed of the exact width and length of one or more frames, a yard and fix inches deep on every fide, sunk half a yard, or more, if dry ground; and at top of which is to be placed the frame with its proper glasses.

In front of the pit, a raised border of rich earth is formed three or four feet wide, in which to plant the trees, taking such as have three or four feet stems, with the heads trained wall-tree fashion, and arrived to a bearing state; planting one for each frame, in the middle of the front, in a slanting manner, for the head to be introduced in the frame towards the glasses; the top of the stem being admitted at an aperture in the bottom of the front of the frame: and a trellis is constructed within a foot from the glasses, ranging parallel thereto, on which to train the branches in the manner of wall-trees.

The pit of this frame is filled with tanner's bark, or hot dung, in February: then put on the glasses; the heat will soon rise, will set the trees in motion, and early in bloom; take care to cover the glasses in cold

cold nights with garden mats; give fresh air every warm sunny day, and sometimes water the branches of the trees when not in blossom, with the same precaution as advised in the other forcing frames; and the trees will furnish ripe fruit a month or six weeks before the common season.

If the bark declines its heat in six weeks, or two months, fork it up from the bottom, mixing the parts well, which will revive the heat, and continue it more effectually till May; though if it is worked chiefly with dung, and the heat is declined, it must be renewed, either by lining the outside of the pit and frame behind, substantially with hot dung, and by removing the worn-out from the pit within, adding a quantity of fresh hot dung, working up the remaining old and new together.

If sharp frosts happen while this kind of frame is in forcing, lay some dry litter around the outside of the frame, to preserve the internal warmth.

A Forcing-pit *by bark-heat*, is a simp'e detached frame, formed in the open air, close on every side, with glasses only at top; it is about six or eight feet wide, and of any length required, formed with nine-inch brick walling, five or six feet deep in the back, to the bottom, by four in front, both ends corresponding, sunk partly in the ground, if not wet, or let about one half remain above the surface, if in danger of wet below, and is sometimes formed with a fire-place, and flues, for occasional fire-heat, the flues carried along the upper part of the back-wall, in two returns, and may also be continued round the ends and front, above where the top of the bark-bed will come, allowing for it to be full three feet depth of bark; a plate of timber being framed around upon the top of the surrounding-wall, from which let cross-bars be ranged to the top of the back part for support of

the

the lights, which muſt ſlide up and down eaſily.

This kind of frame being filled up a yard deep at leaſt, with new bark, early in ſpring, January, February, or March, as occaſion may require, will prove of ſingular uſe in forwarding cuttings of tender exotics, and others that do not reſt freely in the natural ground; and facilitating the freſh rooting of any newly potted tender plants, &c.

Another kind of bark forcing pit is formed entirely of wood, by poſt and planking five feet wide, as much in depth behind, to the bottom, by four in front, properly framed round at top, on which to receive ſliding glaſſes as in the other.

Sometimes a boarded forcing pit, is conſtructed all of an equal depth; that is, only a yard deep on every ſide, and exactly the width of a common hot bed frame. The pit being occaſionally filled with bark, &c. and defended with the frame and glaſſes;

glasses, may be employed to much advantage.

Any of the above described *Bark-heat Forcing-pits*, must be filled with bark a yard deep, or with hot dung, and a foot of bark at top, as observed before; and when the beds become warm, plunge in the pots of plants, roots or seeds.

If you have any quite dwarf cherries, &c. in pots, you may try to fruit them in these pits; pots of strawberries placed towards the front will succeed.

Here you may also place many curious bulbous flowers in pots; such as the *Polyanthus*, *Narcissus*, *Incomparabilis*, and other choicer *Narcissus* kinds, *early Tulips*, *Jonquils*, *Hyacinths*, *Persian Iris*, *Crocus Vernalis*, *Ranunculus*, *Anemone*, *Onothogalums*, *Tube-roses*, *Pancratium-lily*, *Jacobæa-lily*, *Mexican-lily*, and any other spring and summer flowering bulbs.

Also fibrous-rooted perennial flowers, such as Pinks, Carnations, or any other choice

choice sorts; dwarf roses, and other small curious flowering shrubs.

Seeds of oranges, citrons and lemons, when intended as stocks for budding upon, in order to raise trees, if sowed in pots, and plunged in the bark, will soon come up, and effect a quick growth.

A sort of close bark-pit is sometimes formed two feet deep, either planked, or four inch bricked, of the dimension of a wide hot-bed frame, for forcing early peas and kidney-beans; filling the pit with bark in January or beginning of February, and puting on the frame, cover the bark six inches with light mold, to receive the seed or plants.

Pease may be sowed in December or January, either to remain, or for transplanting; chusing the early dwarf sorts. If designed to remain, let them be sowed in drills, an inch deep; the rows fifteen inches asunder; or, if intended to raise the plants in readiness to plant out when the

the bark-bed is made, they may be sowed a month or six weeks before, either in the natural earth, in a warm border, or in a gentle hot-bed; or may be sown thickly, in November or December, either in large pots, or in the full ground, under a common garden frame, for transplantation, giving them the shelter of the glasses, if severe frost or sharp cold weather should happen, and they will be ready for transplanting in January or February; or if those in the pots do not rise naturally before Christmas, plunge them in a dung hot-bed, to run them up; or some may be sowed about Christmas, or soon after, in a hot-bed, or in pots plunged in one; the plants will soon come up; give them abundance of air in fine days; and when they are an inch or two high, having then the bark-bed, &c. ready, and covered with one or more hot-bed frames, and earthed six inches deep with light rich mold, take up all the plants, with all the roots and

and earth possible, plant them in rows in the earth, from the back to the front, fifteen inches asunder, and one inch distant in the rows; giving directly a moderate watering to settle the earth to their roots, and prepare them for rooting quickly and soon flourishing.

They must here be constantly defended with the glasses, but indulged with fresh air at all opportunities in fine mild days, by shoving down or tilting the lights more or less, according to the weather, giving also moderate refreshments of water; and, as the heat of the season encreases, give larger supplies of fresh air, and water likewise; when the plants have advanced some inches in growth, add some fresh earth to both sides of their stems, and when they advance to blossom, be sure to admit plenty of free air; and in fine warm sunny days, shove the glasses almost entirely down, repeating also the waterings more abundantly when in bloom and the fruit setting; let them have also the advantage of moderate warm showers of rain, they will thus furnish pods for gathering in March or April, and by having one or two beds in succession, may continue a regular supply of small quantities as rarities, till the time for the natural crops in May.

Kidney Beans Kidney beans may be managed nearly in the same manner, only they will not succeed by previously raising the plants in natural earth, nor must they be sowed earlier than February. Procure for this purpose some of the small early dwarfs, and if to be sowed at once, to remain immediately in the bed, let it be earthed six inches with light rich earth; then draw shallow drills an inch deep, and fifteen or eighteen inches distance, place the beans one or two inches apart in each drill, they may be sowed in large pots finally to remain, four or five in each, an inch deep; plunge the pots in the bark-bed without having any earth

at

at top, or the plants may be first raised in *Kidney* any hot-bed or hot-house, and being sowed *beans* a fortnight or three weeks before you make the bark-bed for their final reception, either thick in large pits to plant separately, at an inch high, or in small pots, three or four beans in each, and when the plants are two or three inches high, turn them out with the balls about their roots, either into larger pots to remain, or fully into the bed furnished with earth at top for their reception. Let each planting have directly a moderate watering, but such beans as are not come up, must be very sparingly watered, much moisture would rot them; let them be always defended with the glasses as advised for the pease, indulge them with fresh air in fine mild days, and give moderate waterings as the earth becomes dry.

Be careful as the plants advance in growth, to apply fresh earth about their stems, to promote a strong and plenteous bloom,

bloom, and an abundant crop will come to perfection in April; and by two different beds, you may continue the succession to May and June, till succeeded by the early crops on warm borders.

For other particulars, see *Dung-heat Forcing-frames*.

Melons and cucumbers may also be raised to perfection in the afore-mentioned bark pits, under common hot-bed frames; the moderate steady durable heat of the bark bed, is peculiarly calculated for the culture of the Melon, not making the beds earlier than February for either sort; the plants to be previously raised from seed, sowed a month before in a dung hot-bed; sowing the seeds either in a pot plunged in the hot-bed, or in the earth of the bed, they will come up in two or three days.

In default of plants, you may sow seed at once in the bark, in the places where the plants are to remain, some under each glass,

glass, which in beds, not made till February *Melon* or March, may succeed very well, but will not arrive so soon to perfection by a month.

Sometimes a bark-bed is formed with the waste or cast off bark from old hot-houses in March or April, for Melons and Cucumbers; which, if not quite exhausted to an earthy substance, and not long cast out of the bark-pit, being well worked up in a bed, recovers a moderate warmth.

Keep the beds constantly defended with the glasses, and these covered every night with mats till May, admit air moderately, by raising the lights a little, and give moderate waterings; let the glasses be always shut down early towards evening with mats, soon after sun set, uncovering them in the morning about sun rising, or soon after.

In sharp cold weather, defend the outside of the pit and frame, with dry warm litter, or bank up with earth half a foot thick

thick to the top of the pit, to resist the penetration of the frost, and to keep in the heat, or if the heat by accident decays considerably, or that the weather is extremely sharp, renew the heat by a lining of hot dung around the outside of the pit and frame; the back first, then the front and ends, at a fortnight interval; being careful that the rank vapour of the fresh dung does not penetrate internally, which would prove pernicious to those tender plants. To prevent which as much as possible, lay a stratum of earth at top of the lining, to keep down the steam and heat.

Give fresh air every fine day, by raising the upper end of the light, proportionally to the heat of the bed and power of the sun, and according to the early or advanced period of the season, from nine or ten in the sunshine mornings, in the early season, till two, three or four in the evening; and in the advanced season, from seven, eight or nine in the morning, till sun-set, being very attentive

tentive to shutting them close down, either when the weather over clouds, changes sharply cold, or the wind is strong; then hang a garden mat before the opening, where the air is admitted, to prevent the cutting air or winds, from rushing immediately upon the plants; but shut them down always close an hour or more before sun set, in the early season: when more advanced, and the plants stronger, admit more air, both earlier and later in the day.

Melons

Refresh the plants occasionally with light waterings in sunny mornings, from nine or ten, to eleven or twelve, as you see occasion, by the earth becoming dry; having particular attention not to over-water, but always with great moderation; nor too frequent, especially the *melons*, when they are setting, or when the fruit is advanced to full size.

When the plants begin to make an effort to push their first shoots or runners, by forming a bud in the center, stop or

pinch

Melons pinch them at the first joint, which is effected by taking off the first or second bud, rising in the bosom of the second rough leaf, with the finger and thumb, or with the point of a knife: this strengthens the plants, making them grow robust, and bottom well, so as that they will more strongly push out lateral runners to fill the frame; and which generally proves more fruitful than the first would if permitted to run; as the runners or vines of the plants advance, lay them out regular, and peg them down to the earth with small hooked sticks.

Cucumbers When the plants shew fruit, particularly the cucumbers, set or impregnate the fruitful or female flower, with the male blossom, as directed under the article *Fire-heat Forcing-frame*, which must, on no account be omitted.

The fruitful or female blossoms are easily distinguished from the male; the former always discover the germen or embryo

bryo fruit, situated immediately under the base of the flower, very conspicuously, before the blossom expands; and the male flower, erroneously called false blossom, have no germen or apppearance of any fruit underneath; which male flowers are often by the inexperienced, all pulled off, but this is wrong; for, if all are removed, the female or fruit blossom would prove sterile for want of the fecundating powder of the male; you should therefore only thin them where they arise in thick clusters, and remove all in general, according as they close up and begin to decay.

Cucumbers from the time the fruit sets will arrive to perfection to cut young, in a week, or a fortnight, according to the growth of the plants and good state of the hot-beds.

But melons from the first setting of the fruit will be five or six weeks, and sometimes two months before they arrive at full maturity of ripeness.

Having

Having remarked that for melons, a bark-bed made with old waste bark lately cast out of a pine house or stove, succeeds very well for middle crops, formed either in a common forcing bark-pit, or in a boarded one; in February, March, or April, have a good quantity of old bark thrown together in a large heap, a week or fortnight, to promote fermentation, then make it into a bed, and it will recover a moderate heat, but more effectually, if augmented with about one third of new bark; however, in want of this, let cast off bark suffice, making it into a substantial bed, three feet deep; and having some fine stuff at top, set the plants or seed fully in the bark without any earth.

Being furnished with good plants or seed, but ready raised plants of a week, fortnight, or month old, are most eligible, plant fully in this old bark, or with but very little earth; first making shallow bason-fashioned holes, 15 or 18 inches wide,

wide, and six deep, filling them either with fine earth, or some earthy old bark: place two plants in each hole, or in want of plants, sow some seed, in each hole, half an inch deep; and when the plants are a week old, thin out the neateſt, leaving only two or three of the beſt in each part.

As to culture in this place, keep the lights of the frame conſtantly on, with mats at night; and admit air in warm ſunny days as before directed. Water will be required, but very prudently, the bark-bed ſupports a moſt agreeable moiſt heat in itſelf, and conſiderable waterings would haſten its decline: So give only very ſlight waterings in warm days, occaſionally, but ſcarcely any when the fruit is ſitting, nor after it has obtained full growth and beginning to ripen; when water is to be given, let it be at a diſtance from the main head of the plant; for redundant moiſture is not only hurtful to melons in their gene-

ral

ral growth, but more particularly in retarding the embryo fruit from fetting kindly, and the full grown fruit from ripening with a good flavour.

If the heat should become very weak, it may be renewed, by applying hot horse dung against the outside of the pit, first in the back, then by degrees to the front and end, especially when the fruit is setting.

Strawberries may also be fruited early, and in a high state in bark-beds, new and old, half and half, in any common bark-pit, or under a garden frame; the plants being two years old, potted with balls, and placed in such a bed in February and March, giving plenty of fresh air and water, they will fruit in April and May.

Orange trees, lemon trees, &c. are expeditiously rooted in bark-beds, in wide glass departments.

In the spring, great quantities of orange, lemon, and citron trees, &c. are imported hither from abroad, previously raised with

tall

tall large stems, from one to three inches or more, diameter, and from two or three to six feet length, budded at top a year or two before they are sent; and when furnished with strong shoots for the head, they take them up, trim the long shoots, and the roots, and send them here in bundles, without any earth about the roots at all, only wrapped up in moss or the like. Being planted soon after they arrive, in pots, or tubs, according to their size, and plunged in the bark-bed, they strike root, and form as good trees, with handsome heads, in about two years, as we can raise here from the beginning, in ten, fifteen, or twenty; for which reason many persons furnish themselves at the Italian warehouses in London, &c. at from three or four shillings to a guinea each.

They commonly arrive in the spring, and the sooner they are procured and planted the better they will succeed.

Place

Place the root end in a tub of water, for a day or two, then wash the roots and stems clean, trim off the damaged parts of the root, and prune the shoots to six, eight, ten inches, or a foot long, less or more, according to their length and substance, in order to force out lateral shoots below, to form a full, compact, regular head, then plant them in large pots, or in tubs. Having for this purpose a proper quantity of rich light earth, blended, if possible, with some good loam and rotten dung, and placing some tiles or stones over the holes at bottom of the pot or tub; put in some earth, then plant one tree in each pot or tub, fill up with more earth, and give a watering to settle the earth close to the roots; place it directly in a bark-bed of a brisk heat, plunging down to the rims, and they will soon strike root, and begin to shoot at top, and form tolerable good heads by the end of summer, they are

to

to remain here, until the middle or end of July. Wrap a hay-band round the stem, lightly, just to prevent the sun drying the outer bark too much, and let them be well supplied with water, but always with moderation; likewise on very scorching sunny days a slight shade is advisable; and when the trees begin to shoot, let air be freely admitted, and now and then water over the stem and head of the tree, as well as on the earth in the pots, &c. which must not be suffered to get dry, but always be kept moist.

If they produce the shoots thin or stragling, it is advisable in June to prune them to a few eyes, to force out lower collaterals the same year.

In the middle or latter end of July, remove them in their pots into the full air in a defended situation, and supply them with plenty of water during the dry weather; here to remain until October, then to be carried into the green-house.

For want of a forcing-frame department, in which, to strike these trees, it is sometimes effected by plunging them in a common bark-pit, or any temporary one just for that purpose, contriving to erect some garden frames with the glasses, at a proper height to defend hot-beds, and the heads of the trees, and draw them into shoots; closing up the vacancy at bottom with mats, &c. or, instead of frames and glasses, you may defend them with an awning covered at top and front with oiled paper, formed as directed for *oiled paper frames*. Or, for want of better, have a bark-bed formed in a kind of temporary boarded pit in the green-house or any glass-case, in which to plunge the trees.

New budded young orange trees may be assisted greatly in striking, by plunging them in August, as soon as the budding is performed in the bark bed for a fortnight or three weeks; the buds from the sun, and the glasses shaded in the middle

of

of the day, which will facilitate the union of the bud with the ſtock, giving the plants air, day and night, and removing them again into the open air, or into the green-houſe, with the windows all open till the end of Summer.

HOT-HOUSE OR STOVE,

HOT-HOUSE Departments, commonly called Stoves, are designed principally for the culture of the pine-apple, and other tender exotics of the Torid Zone, that require conſtant aid of artificial heat all the year, and that of actual fire in the winter, and in the early part of the ſpring. As theſe departments ſerve alſo to force to early maturity ſome of our choice fruits, flowers, and eſculent plants, with very little additional trouble, the hot-houſe ſhould be both of capacious width to admit of a large bark-pit within, and have brick walls erected ſo as to allow fire-places and flues, from November until April or May, to ſupply more effectually

the

the constant temperature of heat required; which is about 70 degrees of the thermometer; a curious mathematical instrument for measuring the degrees of heat, and with which every such department should be furnished, as a sure guide to regulate the heat in the proper degree.

Hot-house Departments prove also singularly useful for forwarding to early perfection many of our choice plants, fruits, and flowers, facilitating the growth of tender seeds and cuttings, and vines planted in a border without side of the front, and trained into the house.

This building should stand full to the south sun, arranging lengthways east and west, fronted wholly with glass-work.

The dimensions should be twelve or fourteen feet wide in the clear, to admit of a bark-pit withinside, six or eight feet wide, and an eighteen inch, or two feet alley all around, both sides and ends, for a passage. The length may be from twenty to forty, sixty, or an hundred feet, &c. The height ten or twelve feet in the back-wall, by only two or three in front and ends, on which to erect the upright glasses head high, or about six feet, wall and glass work together: and from the top of the glasses, the inclined or sloping lights are ranged up to the top of the back-wall.

The walls should be a brick and half thick, carrying on the fire-place and flues as you advance with the walls, the furnace for the fire-place being formed at the bottom without side, at either or both ends, according to the length of the house. If but thirty or forty feet, one fire may be sufficient; but, if considerably longer, two furnaces will be requisite, each having its proper set of flues, eight or nine feet wide, proceeding within side along the ends and front, thence to the back-wall, extending the whole length in three or more returns, ranging one over the other, separated by broad paving-tiles, the uppermost flue running

ning into the vent or chimney at one end behind; or at both ends, if two fire-places.

The pit or cavity for the bark-bed, is to be six or eight feet in width, by three and an half deep, continued along the middle space, the length of the department, allowing a width of two feet on each side, and at the end, for a path, to pass quite round: it may be a foot or half a yard sunk, the rest raised; formed by a four or nine-inch brick wall, surrounding the cavity, three feet six inches deep from the bottom of the pit to the top; where you should finish with a plate of timber to secure the whole.

Some have the flues carried up distinct from the walls, detached about two or three inches, to have the advantage of all the heat from both sides, without losing any on the outer walls.

Some have also flues continued round the outside of the parapat wall of the

Hot bark-pit, in order to have as long an extent of flues as possible, to obtain the greater benefit of the heat, without wasting the fuel ineffectually; but it must be detached an inch or two from the pit-wall.

The glass-work in front and at top, to inclose the whole internal space, is formed upright, in front five or six feet high, and inclined or sloping over head; the upright range is to be erected along the top of the front wall, three or four feet high or more, to proper framing of wood-work, with the glasses a yard and six inches wide, made to slide; sloping sashes of the same width, ranging from the top of the front to that of the back wall, in two tiers, sliding one over the other, and supported upon stout rafters or bearers, extending from the back to the front, at a proper distance to receive the glasses: but sometimes, in wide departments, there is a third or short upper tier, a yard length, fixed;

fixed; the two lower tiers to slide, the upper of which to slide under the fixed glasses; the glaziers work should be performed imbrication manner, or with the panes or squares of glass lapped, the ends over one another, like the tiles of an house, especially the top sashes, the more effectually to discharge the falling wet, and should be laid in putty, not lead; though the uprights in front may be glazed in the common way, either laid in wood or in lead work: but they are not so effectually water-tight, as in the former manner.

When designed principally for the culture of pine apples, the main stove or hot-house should have the assistance of a smaller stove department, denominated a Succession House, or sometimes only simply a pit. These are of utility in rearing the young pines from the beginning, and forwarding them a year or two, till arrived to a bearing state, then to be removed to the main house to fruit.

The Succession House is sometimes formed nearly on the plan of the main stove, but generally of considerably smaller dimensions, especially both lower and narrower, and sometimes to save width has no alley in front, but you continue the bark-pit quite home to the front wall; furnished with flues for fire in winter, as in the large house.

A Succession Department is sometimes formed in the manner of a pit, as described under the article *Bark-heat Forcing-pits*; being formed in the open air distinct from the other stoves, six or eight feet wide, inclosed with a brick wall, no upright glasses in front, only sliding lights at top; having the back-wall six feet deep by four or five in front; and the whole internal space forming a cavity for a bark-pit, three feet six inches deep; having flues ranging round the walls within, for fires in winter.

But

But when there are considerable collections of pines, they, besides the main stove, and Succession House, have also a still smaller department, called a Bark-pit, as a nursery frame for the young plants of the year; this is formed in the full air, five or six feet wide, having a nine inch wall, five feet behind by four in front, with sliding lights at top; sometimes with one or two ranges of flues along the upper part of the back-wall, to make occasional fires in winter, if the young plants should be continued there during that season.

The general process of these different departments is as follows: the smaller ones receive the young plants, crowns and suckers of the year, in Autumn, to strike and nurse them two or three months, or during the Winter.

They are then moved to the Succession House for a year; after which being two year-plants, arrived to a proper age and strength for fruiting, they are then removed to the fruiting-house or main stove, in September or October.

The use of having smaller stove or pit departments, assistants to the main hothouse for the culture of pines is this; that it not only affords more room in the main stove, to cultivate a larger supply of fruiting plants, without crowding the pit; but the full grown plants succeed by a stronger temperature of fire-heat, to forward them into fruit in due time in the spring, than is generally thought proper for the infant and succession plants; which, if heated too considerably, are apt to fly into small trifling fruit, in Winter or Spring, a year or two before they attain the proper size: whereby you both lose the advantage of obtaining pine apples in due perfection; and of having a supply of succession plants to furnish the fruiting-house the insuing year, for young untimely fruit

fruit-plants prove useless for any future purpose.

By having distinct departments for the yearling and succession pines, we can more readily support such a medium of regular heat, peculiar to the young plants, and forward them properly in growth, without forcing them into fruit: to effect this, you must always keep the fire-heat in winter of a moderate temperature, and always very regular, never to raise the thermometer above the pine-apple standard, rather always under than over; and the bark heat also equal: by no means plunging the pots finally, while the heat is very strong to force them too fast in growth: to prevent which the bark-bed is sometimes formed of only one half new, and the rest old bark of former beds, worked up together, to effect a moderate heat; nor should the bed in which the plants are plunged be permitted to get too weak before you fork it over, and add a portion of new bark, when the old is exceedingly declined; for a want of heat would, on the other hand, stunt the plants; and, by either the extreme of heat or cold, the young pine plants are almost certain to run up to fruit unseasonably; but by observing due moderation, both of fire and bark-heat, you may always have a plentiful supply of proper plants in two or three stages: the crowns and suckers of the year, year old and two years plants, besides the fruiting plants in the main house.

The preparation of the pine-apple, as observed above, is effected by the suckers produced from the bottom part of the old plant, issuing from between the leaves; also by the crown or leafy part at top of the ripe fruit, as well as by small suckers emitted from the base of it; all of which are generally fit towards the end of summer and autumn, when the fruit is ripe, from July till October.

As

At the proper time let them be detached from the mother plants and fruit, and laid in any dry part of the stove or elsewhere, to heal over the moist part that adhered to the plant, that it may not rot by the moisture thereof, when planted. Proceed to plant them in small pots (48s*) of rich loamy earth; previously stripping off some of the bottom leaves next the root part, to give liberty for the emission of fibres, which done, plant one sucker or crown in each pot, filled with light rich mould, pressing the earth closely around them; directly let them be plunged into any warm bark-bed, or in a nursery pit already described, and previously filled either with bark, arrived to a proper heat of a moderate temperature; or hot dung, having some bark of any kind at top six or eight inches deep; or in want of bark, saw-dust, the same depth, in which to plunge the pots; placing the taller plants in rows, behind, and the lowest forward;

giving

giving a very moderate watering to the earth, not over the plants, but just to settle them firmly in the earth of the pot, and facilitate their rooting, which they commonly effect in a week's time, and begin to grow.

As we have before noticed, these young plants may be continued in any bark-pit several months, if you are in want of room in the larger departments.

Observe, as the cold season advances, to make a moderate fire at night; if you are not furnished with a fire place in severe weather, line the outside of the pit, &c. round with hot-dung, especially if the heat of the bark decreases; and if a wooden frame and pit, line up the sides of the frame also with dry litter, to keep out the frost, and to retain the heat.

Observe likewise during the winter, to give only a moderate portion of air in fine warm sunny days, by sliding some of the glasses an inch or two down; water must be be used with great moderation and caution, all winter, not above once a week or fortnight, very lightly on the earth of the pot only, not over the leaves at this season. Cover the glasses every night with thick garden-mats, or in very severe frosts, with litter also, if it shall seem necessary; but as the spring and power of the sun advances, encrease the portion of fresh air, as also of watering, but always very moderate.

Then as to spring and summer's culture, observe the necessary precautions of preserving the heat by the rules advised above; and of admitting fresh air and giving occasional light waterings, in fine, warm days; for we must still keep up the heat moderately during the spring and summer months, till September, this being the principal growing season of the plants; therefore about April they will require to have the bark forked up and mixed with about one third of new, removing first away

the

the moſt earthy parts of the old at top and ſides, then work up the remaining old and new together, and directly re-plunge the pots.

If any have been wintered in dung hotbeds, either in a pit, or in beds in the open ground, that heat will alſo require renewing.

According as the plants in general have advanced pretty conſiderably in growth, they muſt be ſhifted into larger pots in proportion to their ſize, which will be required twice a year, ſpring and autumn; April or early in May for the ſpring ſhifting, and Auguſt and beginning of September for autumn; adapting the ſize of the pot to that of the plants; the firſt ſhifting may be from 48s. or 64s. into 32s. thence into 24s. or ſixteens finally to remain. At each ſhifting be careful to have good freſh loamy compoſt, or any rich light loamy ſoil, or good garden mould, for the new pots, in which, previouſly place pieces

of

of tile, or oyster-shell over the holes at bottom, then some fresh earth; and having the pine-plants ready, proceed and shift them, pot and pot at a time, turning each plant out of its present pot, with the ball of earth entire about the roots, placing it therewith in the new pot, filling up around with more new earth, and finish with a moderate watering; then observing at each time of shifting to have the bark bed ready forked over, adding some new if wanted; but particularly the spring shifting, and directly re-plunge the pots in the bark.

In the operation of shifting, you will observe if any plant is in a weak sickly state, it is proper to detach most or all of the old earth from the roots: examine them, if any diseased or decayed parts appear, either of the fibres or main stock, cut them off and wash all the rotten part well in water, then put them in fresh mould,

mould, give water, and plunge them in the bark.

As the young plants in their proper pots advance in growth, they muſt be taken up and replunged thinner, and ſome removed into other departments; that if they are crowded conſiderably in any nurſery pit, the largeſt ſhould be removed into the next larger department, ſuch as the ſucceſſion houſe, or occaſionally into the main houſe, if you have no other, and that there is room; but where there is a ſucceſſion pit, &c. by no means crowd the fruiting plants in the main ſtove, with others not yet arrived to bearing, but continue them always in the nurſery departments, if poſſible, till the ſecond autumn; then being of good ſize, remove the general ſupply of large plants into the fruiting, which ſhould every autumn, at the end of September, or in October, be filled with a ſucceſſion of new fruiting plants.

Where

When there is only one common stove or pine-house, both to raise the plants from the beginning, and to rear them to a fruiting state, the whole may be effected within the same department, though not always with equal success. Being provided with a proper supply of fruiting plants, at first setting off, and succession ones for the succeeding years fruit; then the crowns and suckers they produce, being potted in small pots, may be plunged in the bark-bed as close together as possible, or stowed between the pots of large plants whenever there is any room, so to take up but as little space as possible, to admit of proper room for the supply of fruiting plants, and as these young ones advance in growth, shift them into larger pots.

In this general hot-house for both old and young plants, be always careful during the winter season, not to force considerably with too great fire-heat, lest it drive all the young crowns, suckers, and succession plants in to fruit; you should be particularly careful to have the fires moderate and equal, so as never to raise the thermometer above the degree marked *ananas*.

Though even in this case, of having only one stove department, and that greatly crowded with young plants, they may be moved out in summer, from April or May, until October or November, in a substantial dung-heat bed, under a common hot bed frame, the bed being first reduced to a moderate heat, having old or new bark at top, or into a boarded bark-pit, of a yard deep, fitted to the width and length of one or more of the above-mentioned hot-bed frames filled with bark; and in either of which nursery-beds, plunge the pots of young pines, out of the main stove, to give room to the fruiting plants, and that the young ones may also have more room to grow. In October or November, the stove will be thinned

ned from the present year's fruiting plants, which may be moved therein for the winter.

With regard to the general management of the hot-house or stove, take the following observations.

The bark-pit of the stove must be continued full of tanner's bark the year round, and fires kept in winter and the early part of the spring; the bark-bed all the year in some stoves being necessary to effect a constant regular heat in all seasons, and in which to plunge the pots of pine plants always to remain; for they will not succeed or fruit well, without being placed with their roots in the pots, in the kindly moist warmth of the bark-bed, which is peculiarly adapted to the growth and fruiting of these tender plants.

The hot-house and other pine-apple departments should have the bark-pit filled with new tan or bark, every autumn season about October, when the old beds are pretty well cleared from the other

plants

plants that produced the fame year's pineapples; and at which time, after having refreshed the bark-bed, that of the main house must be filled with a supply of new fruiting plants of two years old from the succession house, this furnished with yearling plants from the nursery-pits, and these with young crowns and suckers of the year, taken new from the old mother plants for propagation, to support the proper stock, in three different stages of growth one under another.

Observe also of the yearling and succession plants, they should likewise have the bark-bed, &c. renewed in October, either wholly, or half new, worked up with the old, plunging the pots of plants in the bark, &c. with the precautions and regularity advised for the fruiting plants.

Any other very tender exotics that require bark-heat to support their proper growth, should also be now plunged in the bark-bed, such as the sensitive plants, &c.

&c. though moſt ſorts, except the pines, will ſucceed, placed any where in the ſtove, upon ſhelves, and the ſucculent plants, being of a fleſhy moiſt temperature, may be placed on the top of the flue, and other dry parts of the houſe.

The bark-bed formed of entire new bark in October, will keep a proper temperature of heat, till January or February; when, if it is much declined, take up all the pots, and fork the bark over to the bottom, plunging the pots again directly, and this operation will renew the heat till March or April, at which time an additional ſupply of about one third or more of new bark will be required; firſt removing all the worn out earthy ſtuff at top and ſides, a foot down, then fill up with the ſupply of new bark, forking up the whole, new and old together to the bottom, and plunging the pots as before, it will, after this, maintain a good heat all ſummer, till September or October, when, having

H   moſtly

mostly exhausted its fermentative property, it must be almost wholly removed, and the pit filled up with entire new bark, as at first, previously screening the old in the pit, retaining the large, and clearing out all the earthy part that passes through the screen, then adding the new, working it up with the remaining old, raising the whole some inches above the brim of the pit-wall, as it will settle half a foot in two or three weeks. Having fermented to a proper degree of heat, in a week or fortnight, it will be ready to receive the supply of new pine plants in their pots, as before directed; and thus the bark-bed of these departments, for the culture of pines and other tender exotics, must be annually managed.

All Winter, from November until April, or longer, the internal heat must be assisted with fire-heat, of coal, wood, peat, or turf, such as may be most conveniently obtained, every night, from sun-set till nine

(147)

nine or ten o'clock, to maintain a proper heat till day light; in cold, or very raw foggy mornings, make also a moderate fire, and in severe weather continue it with moderation all day, or every other day, as it shall seem necessary; being directed by the thermometer, which should be continued always in the stove, as your guide; it having the degrees of heat for pine-apple and other plants of similar temperature, generally marked *ananas*, and the fire heat should be regulated accordingly, to keep the spirit of the thermometer within, about $5°$ over or under. Before Christmas, keep the fires more moderate, but afterwards, by gradual increase, a little brisker, in order to forward pines into fruit, in February and March, still however observing the thermometer as your regulator.

Likewise during the Winter, keep fires in such of the succession pine-apple departments as are furnished with flues, but keep them more moderate than in the fruiting houfe, for fear of forcing them into fruit a year too foon.

In rigorous frofty weather cover the glaffes at top, either with large garden mats, or extenfive pain ed canvas cloths, or rolls, to let down and pull up by pullies and lines; which occafional canvas is to be ufed only in fevere frofty nights, nailing mats alfo againft the front glaffes; and if the froft continue remarkably intenfe, and no fun, the top covering may be continued occafionally in the day time, but only during the very fevere weather.

Admit frefh air occafionally, both in Winter and Summer; in Winter with caution, in fun-fhine calm days, from nine, ten, or eleven in the morning, till two or three in the afternoon, according to the power of the fun, and external heat of the ftove, &c. by fliding fome of the front glaffes a little way open; being fure to fhut clofe in due time in the afternoon,

noon, or sooner, if the weather changes cloudy, or the air blows sharp. As the warm weather comes on, give air more freely in all sunny days; and when the sun is vehement, freely draw open the glasses.

Occasional waterings are necessary at all times of the year, not however above once a week in Winter, just to keep the earth in the pots moderately moist, but not to water over the leaves at this season. In the advanced warmer season give water freely, and in the heat of summer, two or three times a week.

Watering must also be performed to all other plants in the hot house; those plunged in the bark, will not require it so frequent or abundant, as those placed in the open space, upon the shelves, flues, &c. the woody kinds will require it most, the herbaceous sorts must also have a proper share; the succulent plants require it least, but these must not be neglected.

The old pine-apple plants, after having once borne fruit, produce no more, but they furnish off-set suckers, as before remarked.

The crown or top of the ripe fruit is separated when served at table, not before; and from these the plants are encreased, and the species continued. If any old plants, after the fruit is cut, furnish no off-sets, cut down the leaves near the bottom, and plunge them in any hot-bed, or bark-bed, where there is a brisk heat, which will facilitate their emitting suckers, which are to be detached, and planted in small pots plunged in any hot-bed, and managed according to the former directions; and by thus propagating a sufficient quantity every year, you will be supplied with a constant succession of fruiting plants; both suckers and crowns, requiring generally two years to grow to a proper size, to produce good fruit.

In

In the hot house the various other exotics kept for variety being potted, they may be placed wherever there is room, and shifted occasionally into larger pots and fresh earth, once in a year or two, as they shall require.

Kidney-beans, cucumbers, strawberries, &c. by being placed in pots, and disposed in the house, will succeed abundantly well in this degree of heat, and may be continued in succession till the season of their natural growth. For particulars in their management, observe what is directed under the articles *Fire-heat* and *Bark-heat Forcing frames*; the cucumber has generally the best chance in the pine-stove; having contrived to raise the plants previously in any dung hot-bed, &c. in small pots, and when in the large rough leaves, turned out with balls into large pots or boxes as before described.

In the hot-house, you may have pots of flowers, of the bulbous and fibrous-rooted kinds,

kinds, placed upon shelves, plunged in the bark-bed: Likewise feeds of tender annual flowers, such as balsams, globe-amaranthus, and the like, sowed in pots in February and March, and plunged in the bark bed, will grow freely, and when an inch or two high, prick them separately in small pots, giving water, and plunge them also in the bark, supplying them with water, and as they grow large, shift them into larger pots. Any other early flowering kinds, such as ten-week stocks, &c. may be soon raised to a handsome flowering-state.

The hot-house proves exceedingly convenient, in striking cuttings of various tender exotics, both green-house and stove-kinds; planting them many together in pots, and plunging them in the bark-bed. Also new potted plants having bad roots, or such as you would run off expeditiously, by taking fresh root, which may be effected, ed, by plunging them in the bark-bed, for two or three weeks.

Pots of choice flowering shrubs of small growth may be placed here occasionally in Winter and Spring, to forward them early into flower, such as roses, syringas, honeysuckles, cistus, oleanders, oranges, hypericums, &c. being potted and trained with low heads, and placed in any vacant part, they will blow early in tolerable perfection, though they are oftener succefsful in a more moderate heat than this of the pine-stove; however, place them in the most airy situation, towards the front and ends, and they will have the better chance.

By planting vines for training into the hot-house, we have always the opportunity of obtaining early grapes without trouble.

For this purpose, some good three, four, or five year old plants arrived to bearing, should be planted in November close against the front, in any light warm soil, one against each support of the glasses; and being pruned to one strong shoot for a stem, several feet in length, train it in through a hole, two or three feet from the bottom, thence lead up along the main supports or bearers of the front and inclined glasses: afterwards some shoots trained at wide distances under the top lights, and managed as directed in the article, Forcing-houses, Vineries, &c. they will here, without any other care than annual pruning, produce grapes in the greatest perfection, two months before those in the open ground, and with an improved rich flavour. The proper sorts of grapes for the above purpose, are exhibited in the article Forcing-houses, in the Fire-heat departments, &c.

The culture of pine-apples, without the assistance of hot-houses, or fire-heat, is sometimes effected with success.

A bark and dung-heat department, is employed

employed for this purpose, to save the expence of larger erections.

It consists of a pit formed in the open air, six feet wide, and covered with sliding-glasses at top, being either six feet deep behind, by four or five in front, and filled a yard and six inches deep with bark, and the outsides occasionally supplied with hot dung. Or, a pit formed three feet six inches deep on every side, either with a wall of brick, or with strong deal boards; augmented at top with large garden frames, made exactly to fit, and furnished with sliding-lights; the pit being in length sufficient for one, two, three or more such frames, ranging in a line, of inch and half or two inch deal, and three feet six inches in the back, and two and a half in front, for raising the young plants, but for the fruiting plants, four feet and a half, and a yard six inches in front, by ten or twelve feet long each, with three sliding lights at top,

top, three upright ones in front, and one at each end.

To have this as convenient as poffible, one frame or a fet of frames might be contrived to ferve a double portion of pit; the frames made to flide along from one divifion to another; as the heat in the firft declines, make a frefh bed in the referved pit, then fhove the frame along over it, fo moving the plants in their pots accordingly; which will be particularly convenient when dung heat is intended, in which three or four new beds will be required to continue the plants the year round.

If intended to work with dung entirely, it may be affifted without any pit at all, making the beds on level ground, and defended with the frames; four different new beds will be required, as aforefaid, to carry the plants on through the year, and each bed fupported with occafional linings.

But

But to have every thing as convenient as poſſible, it is proper to have both a nurſery-frame, and a fruiting frame; that is the nurſery frame to ſtrike and raiſe the young plants, and the other to receive the large plants to fruit them; but the nurſery frame need not be ſo deep by a foot, or half a yard, as the fruiting frame.

Let the pit for theſe frames be formed with nine inch brick work, or ſtrong two-inch planking, nailed to proper poſts fixed in the ground; and either ſunk half a yard, more or leſs, according as the ſoil is wet or dry; obſerving, if to be worked chiefly with dung, a boarded pit is rather the beſt, throwing in the heat more effectually; or even, if for bark heat, the ſame kind of pit is very proper, in order to receive the aſſiſtance of occaſional dunging externally; for the ſame purpoſe, it might be brick in front and both ends, and boarded behind; or a moveable temporary pit, framed with

ſtrong

strong planking for bark-heat, might be continued.

Have the nursery-frame the same width of the pit, three feet six inches behind, by two and a half in front, and ten and a half long, with three lights at top.

The nusery frame for the reception of the crowns and suckers of the year, will be required in August or September, to continue them here one year, or till large enough for fruiting, then to be removed into the fruiting frame; observing, that during their abode in the nursery-frame, they may be supplied with fresh air and occasional waterings, consistent with the rules explained in the hot-house culture, covering the glasses every night with mats; and, as the heat declines, the outsides to be supplied with hot-dung. But a new bed will be required every two or three months during the winter, from October till March or April; thence once in three months till the end of summer, defending
each

each bed with dry litter, or a rampart of dry earth.

The fruiting-frame should have the pit filled with new bark, or dung and bark together, about the end of September or in October. At this time, fill the pit a yard six inches deep, either with entire new bark, as the most eligible for its regular and durable heat the year round. Or if more suitable to your convenience, fill the pit with fresh horse stable-dung, in the manner of making common hot-beds; and in a week or fortnight, when settled about a foot, fill up with bark, new or old, as can be obtained, or with saw-dust, to keep in the heat, and in which to plunge the pots. Put on all the glasses, keeping them close till they draw up the heat, then open a little for the steam to pass away.

If you judge the heat is too great, plunge the pots only a little way at first, and fully to their brims in a week or fortnight after, as observed of the hot-house,

disposing

disposing the tallest plants behind, down to the lowest in front; and directly put on the glasses, and manage as before directed, as to warmth, air and water.

If the hot-bed in the pit is of tanner's bark, line the outside round with dry warm litter in Winter, or raise a wall of earth; but if, in November or December, the back is lined substantially with hot dung, renewing it in a month, it will keep up the heat till January or February; then, if forked up, it will renew the heat and, assisted with a dung lining soon after, support a proper temperature till April; at which time, removing some of the worn-out earthy bark at top and sides, fill up with new, and fork up the whole together; and this will supply an effectual warmth in July, August, and September. In October the pit must be filled with entire new bark, and a succession of new fruiting-plants plunged therein for next year's fruit.

But

Hot houses

But if the hot-bed in the frame is of dung, made in September or October, two new ones besides will be requisite to conduct the plants through the Winter, till April or May, with the assistance of occasional linings to each bed; and another at that time, to carry the fruit on to full growth. If the dung bed is made in a pit, and there is a spare pit, make a new bed therein, about December, in due time to receive the plants in their pots when the first bed is declined; at which time, move the frame along from thence over the new bed, which being covered with bark or saw-dust, plunge the pots; and removing the worn out dung of the first bed, give room to make another at the proper time, about the middle or end of January, or in February, &c.

Observing in the interim, during the Winter, that you defend each bed with litter all round the outside of the pit, and up the sides of the frame; and in a month after it

is

is made, remove the litter behind, and apply a strong lining of hot dung close to the back of the pit, a yard wide at bottom, sloping about two feet upwards, not however raising it the full height at once, but by degrees to the top, and continued along the front and both ends in a fortnight after; covering the top with earth or bark, &c. to keep down the heat, and shoot off the wet.

After this observe, that when the heat of the linings of dung decrease, fork over the old dung to the bottom, casting out the most exhausted, and adding a portion of new in its stead, and work the new and remaining old up together, regularly against the pit and frame as before; the linings will then acquire a fresh fermentation and heat, and renew and support that of the bed for some time longer, When the dung of the linings at any time appears wholly exhausted in respect to heat

heat, it must be renewed with entire new hot dung.

When it is intended to have the fruiting bed made wholly of hot dung, in the open air, upon level ground, without any pit, they should also be made of the dimensions of the fruiting frames, in respect to width and length, and three feet and a half deep of dung, or more, to allow for its settling; and when the frame is put on, lay six or eight inches, or a foot thick of bark, &c. as directed in the pit-beds, to keep in the heat, and in which to plunge the pots of pine plants.

The management of this bed, and plants therein, is nearly the same as directed for that in the pit, both in regard to occasional linings, of hot dung, &c. and of making new beds, as the old ones decline their heat, at the end of every ten or twelve weeks; keeping up the heat during that time with proper linings; and

and give proper supplies of fresh air and water, with the necessary precautions formerly recommended, and cover the glasses every night.

In all these different beds, both those formed in the pit, either with bark or dung, or with dung in the open ground, be careful that the roots of the plants have not too much heat, nor are plunged too hastily into the bed, till you judge the burning heat, if any, has subsided. At the same time do not lose the opportunity of plunging them, as soon, as in due order.

Observe exactly the same methods as directed in the hot-house culture; always keeping up a stock of succession, and fruit-plants, regularly to succeed one another.

DUNG

DUNG-HEAT FORCING,

AND

COMMON HOT-BED FRAMES.

DUNG HEAT FORCING-FRAMES, both in upright fixed erections, and in the common moveable garden-frames, for common dung hot-beds, are employed as the cheapest of all the kinds of forcing-departments; besides many have the advantage of horse-dung of their own, or at a very easy expence.

On these considerations the business of forcing by dung-heat may, to many, be more convenient than any other.

The

The utility of horse-dung for various forcing-works in gardening, is very great, both in the fixed upright fruit, and flower-forcing frames, applied against the back wall, and sometimes formed into a bed, in a pit, within side in more capacious frames in the kitchen garden, covered with common frames and lights, for forcing early cucumbers, melons, asparagus, peas, kidney-beans, sallading, radishes, carrots, &c as also early strawberries, likewise all sorts of curious tender annual flowers, and occasionally any choice bulbous and fibrous rooted perennial flowers.

What we mean by horse-dung for the purpose of forcing, is the dung and wet litter together of the stables: the dungings or droppings of these animals would answer no purpose without the moist litter of the stalls mixed therewith; the standings or stalls being thickly littered down every night, and rendered wet by the dunging and staling of the animal, may

every

every morning be all cleared out to the dung hill, where it ferments and heats, often at first too vehemently for vegetable growth, but by preparation, as we shall direct, is reduced of a proper temperature, and of a more durable heat.

This dung must be fresh, moist, steamy and full of heat, not having been above a month or two in collecting from the stables, that which has lain long, so as to be in a decayed state is improper, which is often the case when it has heated violently, and exhausted its fermentative property; sometimes half rotten and cold. Frequently the very long dry strawy litter, is burned considerably, and has a dry whitish mouldy appearance; which will do nothing in this business, unless well incorporated with that which is fresh and hot; therefore the newer the dung the better, warm and abundantly full of moist, steamy parts, long and short together; or though ever so long and strawey, if moist and warm, by well mixing

ing together, and casting up into an heap, it will acquire proper fermentation for any pupose of forcing or hot beds.

As to the preparation of it, let it be brought in due quantity from the dunghill, then toss it up in an heap, or if a large quantity, in an oblong high ridge, shaking the parts well together.

Then let it remain ten days or a fortnight, or longer, according to its quantity, and quality, that the rank pernicious stench, and noxious burning steam may evaporate. If the quantity is considerable, turn it over once again, which gives greater vent to the noxious vapour to pass off, and the dung to incorporate.

If you are obliged to use dung rather of an exhausted state, let it be improved by the addition of fresh; or if in want of this, or of a sufficient quantity, procure some barrows of coal ashes, in the proportion of two or more to each cart-load of dung, which will promote a more quick

and

lively heat; the dung being caſt up and mixed with the coal-aſhes, and if it be rather dry, throw ſome pots of water over it from time to time, as you advance in toſſing it up in the heap; the aſhes will not only aſſiſt in recovering a briſk and ſtronger heat, but render it alſo of longer duration.

This improvement of the coal aſhes is required principally for weak dung; in that of a ſtrong quality, full of warmth, it might cauſe a too vehement heat. It frequently happens, that aſhes are promiſcuouſly thrown on a dunghill, near the houſe, in which caſe, if they ſeem too abundant, ſhake ſome out, according to diſcretion, in removing the dung.

Thus far is neceſſary to be obſerved in the preparation.

As to the uſe of it, I would remark, that although dung-heat forcing frames, for fruit trees, are not ſo generally effectual and ſuccefsful in the earlieſt forcing, as

fire

fire and bark heat frames, yet they often furnish good fruit a month or six weeks before those in the season. And it is worth the attention of those who are plentifully furnished, to erect some upright frame and glass-work, either against a south wall with trees already formed, or a wall, or strong boarded fence, planted with bearing trees, applying the dung at the proper forcing-season behind, on the outside, &c. as hereafter directed.

Dung heat retains a proper temperature, for this business about six weeks, in which time it will require to be renewed, or the heat revived, and removing away such as is rotted and worn out, and then applying new to the remainder, working the whole up together.

These forcing-frames consist of fruit dung-heat frames, flower dung-heat frames, and common hot-bed frames, both for esculents and flowers; the two former are fixed erections, having upright back walls or boarding,

ing, five or ten feet high, worked principally by the dung placed thickly againſt the outſide, behind, and at both ends; and ſometimes alſo in a pit within ſide; but the common moveable hot-bed frames are worked by the dung, formed into a bed diſtinct from any erection; in all of which the heat is continued by occaſional application of freſh dung, as the heat of the old bed declines, or by removing from the back of the frame the exhauſted part, and adding a quantity of new worked up together as before.

A FRUIT DUNG-HEAT FORCING FRAME, is formed commonly in the manner of a hot-wall, but worked by the hot dung being applied behind, or ſometimes in a pit within-ſide; is a fixed erection, five, ſix, or eight feet high behind, and of any length required.

The dung forcing frame, worked by the dung behind, may be only four or five feet wide, for one row of trees, arranging against the back, and small plants before; having the back either a nine-inch brick-wall, six or eight feet high, or only the foundation of brick work, and strong planking, upward, closely joined, that no steam of the dung can penetrate, the front formed only about six inches or a foot high; from the top of which, to that of the back wall, range the glasses, sloping to about six, ten, or twelve inches width at top, received into proper frame-work; and supported also upon bearers, rafters, or cross-bars, reaching from the top of the back to the front, the width of the lights distant from one another, having the whole well boarded in, water tight.

A frame of the construction designed principally for forcing early roses, and other small plants in pots, may be formed only four or five feet wide, and five or six feet high behind, entirely of wood work in the back and ends, of two inch deal, the ends the same, not above a

foot

foot high in front, and fronted with moveable glasses as above; which frame will blow many sorts of plants to early perfection with little trouble. *See Flower Dung Heat Frame.*

With respect to the fruit dung-heat erection as before described, you will provide some trained trees arrived to a bearing state. The peach, nectarine, apricot, cherry, vine, &c. should be planted in November, in the border, against the front of the back-wall of the frame; and the branches should be regularly trained, either to the wall, or to a trellis, as in the fire and bark heat frames.

As to the time for forcing them, it must not be earlier than the middle or end of January; when being provided with a good quantity of new horse-dung, properly prepared, let it be piled regularly against the back-wall, and ends of the frame withoutside, full two feet wide at bottom; but if three the better, drawing it in gradually

dually upward, to half that width at top, five or six feet high, and according as it settles add more dung, to continue it to the proper height, finishing the top sloping, to throw off the wet; likewise lay some earth at top, beating off smooth with the spade, to keep down the heat.

The lining thus formed will begin to work in a week's time or thereabouts, so as to throw its heat internally, and set the trees on growing, and will continue a good heat a month or six weeks, or longer. As the lining settles down still add more hot dung at top, which will revive and augment the heat for longer duration; but being careful when the general heat decays considerably, to renew it, by forking over and adding new dung; beginning at one end, fork over the old, throw out the rotten and decayed, adding an equal portion of new, working it up together against the frame, as before directed.

If

If you are fearful of losing the internal heat, in the time between removing the old and the working of the new lining, perform only half at a time, proceeding with the middle space first.

Thus the heat of this frame is to be supported, three or four months; repeating the linings twice or more or oftener, as you shall see necessary.

Admit fresh air in all mild days when the sun shines warm, but keep all close in nights and all cold weather, give also occasional waterings to the borders, and sometimes over the branches, according to the precautions observed in the fire-heat frames.

By supporting a constant heat in the manner above directed, together with due supplies of water and free air, the trees will blossom freely, and if it proves a favourable spring, will produce some good early

early fruit, a month or six weeks before the season.

Pots of small plants, as strawberries, flowers, and small roses may be placed towards the front, where they will grow freely, and yield their produce early at an acceptable time.

A *Forcing-frame with a dung-pit* within, may be ten feet wide, with a back-wall, eight or ten high, with upright glasses in front, and inclined lights at top; forming a pit four or five feet wide within, planked or bricked on the sides; and a four-feet border behind, in which plant the trees, to be there trained to the back-wall, and with vines in front, to train up by the glasses; filling the dung-pit with well proportioned hot-dung, early in February, working it in firmly, and regular, as in making a common dung hot-bed, raising it about six inches above the top of the pit, as it will settle a foot or more in a

week

week or fortnight's time; when, if you have any caſt off bark, cover the top of the dung therewith half a foot thick, or with light dry earth, which will keep down the ſteam and the heat from evaporating too faſt, and ſerve alſo to revive pots of ſtrawberries, kidney-beans, and other ſmall plants; and pots of curious ſeedling annual flowers, alſo annuals planted in pots, ſuch as tricolors, cock's-combs, balſamines, globe amaranthus, &c. to forward their growth, and draw them up to a goodly ſtature.

The management of this frame is nearly the ſame as the former dung-heat fruit-frame; admitting freſh air all fine days, with occaſional waterings.

When the heat of the bed decays, it muſt be refreſhed, either by taking out half the old dung at top, and fill up with new dung, worked firmly down, but not trodden, which would prevent its heating regularly; or more effectually to

fork

fork up the old bed wholly to the bottom, first taking out some at one end, forming an opening of a yard wide, for room to work, then proceed to fork over the old dung into the opening, having at the same time a quantity of new ready, and as you proceed in forking over the bed, cast out the worn out stuff, and throw in the new, working it up firmly with the remaining old, sufficiently to fill up the pit as at first, allowing room for a stratum of bark or earth six inches thick, in which to plunge the pots of plants.

Such a frame as above, six or eight feet high behind, by five in front, is sometimes employed as a glass-case drawing frame, in which to draw the choice annual flowers to a tall stature, such as the *amaranthus tricolor, celosia cristata* or cock's-comb, they being previously trained from seed in a common hot-bed, and forwarded in pots to two or three feet height, then removed into the drawing frame and plunged

plunged in the hot-bed made in a pit, as before explained; giving plenty of water they will run up quickly to a handsome size.

A Flower dung-heat Forcing-Frame, is an upright fixed erection, built with a wooden back and ends, and fronted with sloping glass sashes; in dimensions it is four or five feet wide, by any length required five or six feet high behind, formed of strong inch-and-half deal board, well framed together, the front raised up not above a foot; thence are ranged the bearers for the glasses, to the top of the back, supported at about a foot width, from the top of the back erection, having the top of all well weather-boarded to shoot off the wet; within have ranges of shelves rising gradually behind one another, on which to place pots of roses, the tallest behind, and with other lower flowering shrubs in regular gradation to the front; where may also be placed pots of pinks and other low flowers; alfo any choice bulbous rooted flowers, fome in pots, others in water-glaffes.

This frame is to be forced with the dung behind; in January, February or March, having frefh hot dung prepared as advifed on former occafions; lay it up firmly againft the back and ends of the frame, two feet wide or more at bottom, narrowing it gradually in a floping manner to about a foot width at top; in a week's time it will probably be fettled a foot, when more dung muft be added in proportion; laying an inch or two depth of earth upon all, to keep down the heat.

After this, the lining will foon throw in a fine growing heat, and the plants thereby difcover an early vegetation. And being properly fupplied with frefh air in fine days, they will advance in a very agreeable manner; and if fupported in a due warmth, (as frequently directed

will

will flower with great elegance many weeks before their natural season: the bloom may also be continued by a succession of fresh plants, till succeeded by those in the open ground.

This kind of flower forcing-frame is much used among the forcing gardeners in the environs of London, where dung is plentiful, good and easily obtained; and by which, in working the above frames, vast quantities of roses, pinks and other small flowering shrubs, and herbaceous flowering plants, are forwarded to an early bloom, for the supply of Covent Garden and other markets, where we are presented with a variety of the most elegant flowers, some months before their natural season.

Hot-bed Frames for common Dung Hot-beds, being low moveable wooden-frames, are of very considerable utility in almost every part of gardening.

These frames are constructed with inch, or inch-and-half deal smoothly plained;

plained, and are of different dimensions, agreeable to the several purposes for which they are intended; but the most general hot-bed frame for common use, is three yards and a half long, by one and a half wide; two feet depth behind, by twelve or fifteen inches in front; being generally as high again in the back as in front, to give due slope for the glasses, to shoot off wet, and receive the due advantage of the sun; having both ends in proportion, and with these moveable glass lights at top to fit it exactly; but, from a yard to fifteen inches in depth behind, by half that in the front, is the deepest and the shallowest for common frames; though shallow frames of only eighteen inches, or two feet behind, and nine inches or a foot in front; prove the most eligible in early forcing of cucumbers, melons, sallading, strawberries, &c. that by being low, the glasses near the plants, receive a greater benefit

of

of the sun, in proportion, which is a material consideration in early forcing.

But for general service, I should prefer a frame of middling depth, furnished with three glass-lights, made to fit the top completely; that if the frame is ten feet and a half long, each light must consequently be one yard six inches wide, by four and and a half long, being the width of the slope or top declivity, where must be two cross bars ranged the width of the lights, distinct from each end, and from one another, both to strengthen the frame, and for the support of the glasses, which are made to slide up and down, and to move off occasionally.

Besides the common three-light frames, it is proper to have also two others of a shorter length, but of the same width; let one be only for one light, called a one-light box; another with two-lights, denominated a two light frame; which two lesser frames are more particularly requisite in

private

private gardens when but moderate quantities of plants are required, and are necessary to serve as a kind of seed and nursery frames to the larger three-light ones. The one-light frame serves for a seed-bed to raise a few early cucumbers, melons, curious annual flowers and plants, for the first crops; and the two-light frame for a nursery-bed in which to prick the seedling plants from the one-light box, to forward them to a proper size to transplant into the three-light frame.

The deepest frames of about a yard in the back, by half a yard or two feet in the front, are principally either for forcing or drawing frames, sometimes for forcing dwarf fruit-trees in pots, such as Duke Cherries, Dwarf Peaches, Currants and the like for curiosity: also pots of roses and other low flowering shoots, having beautiful or fragrant flowers; and also sometimes for pots of bulbous pinks, &c. placed towards the front; likewise occasionally fionally for drawing curious annual flowers up with tall ſtems, having been previouſly raiſed from ſeed in ſhallower frames.

A DRAWING FRAME for tall annuals, ſuch as tricolors and giant cock's-comb, to receive them in pots from other frames, ſhould be five or ſix feet deep on the back, by three, four, or five in front, and if the front is of glaſs work to admit the ſun, and light, it will be of much advantage; and for which frame a pit muſt be ſunk, in which to make the hot-bed.

But ſometimes a ſort of drawing or multiplying frame for annuals is made in two, three or more different frames of equal length and width, to place one on another, to augment the depth according as the plants riſe in height and require it; and this is commonly called a multiplying frame; being compoſed of ſeparate parts as above; one is about three feet deep in the back, by two in front, furniſhed with three lights at top, two others are eighteen

inches

inches deep on every side, but with no glasses, those of the other frame serves for all; they must be of the same width and length to fit one another so exactly as to appear like one frame; first beginning with the deepest frame furnished with the glasses, and when the plants have grown that height, remove this frame, place one of the others without glasses in its place, either in the same or in a new hot-bed if required; and upon which frame, place the first frame with the lights, and according as the plants advance again to the glasses, add the third and last frame, still placing that with the glasses uppermost, and when all is thus placed, they will, together, form a depth of six feet, which is a competent height for any annuals; but in these very deep frames, it is of great consequence to contrive the whole front of glasses, to admit light to the lower parts of the plants, otherwise they are apt to cast all the lower leaves, discovering their naked stems.

Observe,

Obferve, it is proper that all thefe deep frames have a moderate pit in which to make the hot-bed, funk half a yard or two feet deep, efpecially for the final bed to draw them to their full growth, fo having the bed three feet thick of dung, one half funk in the pit, the other above.

A GLASS-CASE FRAME is alfo very eligible for drawing annuals finally in their laft ftage of growth; being an erection fix or eight feet high behind, and four, five, or fix in the front, by fix or eight wide; upright glaffes in front, and inclined or floping lights at top. Withinfide have a funk pit, four or fix feet wide by two feet and a half deep, to be filled with hot dung, and this to be covered with fome wafte bark or dry earth at top, fix inches deep; and in which the plants, after being forwarded to one, two, or three feet ftature in common hot-beds, are to be plunged in large pots to have their finifhing growth.

In want of proper drawing frames, as above,

above, a temporary one may be contrived with a common hot-bed frame, to force or draw curious annuals, which often anſwers the purpoſe tolerably well.

This is effected by having a ſtout wooden poſt forced in the ground at each corner of the bed, four or five feet high exactly, to receive the four corners of the frame within them; in each poſt holes are bored in the inſide an inch wide, and ſix inches one above the other, in which to place ſtrong wooden pins, one for each poſt; that when the plants in the frame advance ſo much in growth as to touch the glaſſes, raiſe the frame ſix inches, by placing the pins in the poſts, in which to reſt the bottom of the frame; and according as they advance higher, continue raiſing the frame in the ſame manner; and cloſe up the vacancy at bottom with garden mats nailed to the frame, and to the poſt at each corner.

However, with regard to general garden frames

frames for kitchen garden and small flower forcing, they must always be of moderate depth, half a yard or two feet in the back, by nine to twelve or fifteen inches in the front.

The London gardeners who employ a vast quantity of frames for forcing, have most of their kitchen garden frames shallow; for deep frames not only require a greater force of heat, but the glasses being further from the plants they do not receive so great benefit of the sun, in winter and early spring; besides, many sorts of esculent plants of quick growth are apt to draw too fast up in height, where there is a large space between them and the glasses, as they naturally aspire upward to the light; and the more space they have, the more apt they are to run up weak, particularly cucumbers; also early lettuce, and cauliflower plants, wintered in such frames without heat.

For these reasons most of the above gardeners have frames for wintering tender plants, such as lettuces, cauliflowers, sallading and for forcing early raddishes, cucumbers,

cumbers, and the like, not more than half a yard in the back, by nine inches in front, and some only fifteen inches behind, by seven in the fore part.

But frames for forcing early strawberries, kidney-beans, peas, &c. should be half a yard, or two feet, deep behind, by one foot fifteen inches in front.

Nevertheless, deep moveable frames of two feet and a half, or a yard depth in the back, by half a yard or two feet in the front, prove useful in wintering many tender plants in their infant state, both of the hardy kinds that commonly succeed in the open ground, but are tender whilst young, such as the *arbutus*, some kinds of *cistus*, &c. as well as many herbaceous kinds: likewise to winter many of the hardier kinds of small green-house plants, such as myrtles, &c. either for want of a green-house, or a frame of ease where the green house is crowded with plants.

Or, for green-house plants, you may have a frame a yard and a half deep in the back,

back, and a yard in the front, having the front alſo glazed, confiſting of three upright ſliding faſhes, beſides the three ſloping lights at top: admitting a larger ſhare of ſun and light will prove confiderably more beneficial to the plants, than frames having a boarded front.

When deſigned to winter any of the tender kinds of full grown ſhrubs, &c. or green-houſe plants, in either of theſe kinds of deep frames, it is adviſable to have them in ſeparate frames, the full-grown plants in one, and the green-houſe kinds in another; having the plants in pots, place the talleſt behind the lower ones towards the front, according to their ſize; and in winter, when the froſt ſets in, it is proper to line the frame round about with dry long litter, not warm litter to force the plants; or part may be banked up with a rampart of earth, the reſt occaſionally with litter: and the glaſſes covered every cold night with mats, dry ſtraw, &c. eſpecially the green-houſe plants;

obſerving

observing to give abundance of air every mild day, but more freely to the full-grown plants, which should have all the air every day when not severely frosty, by shoving the glasses entirely off.

Likewise a frame of two feet and a half or a yard deep behind, by half the depth in the fore part, is often used as a nursery-frame for receiving young pine-apple plants, particularly the crowns and suckers of the year, and sometimes also the yearling plants: having a strong hot bed, made the size of the frame, in September or October, either of tanner's bark in a pit, or of horse dung and bark together on level ground, or entirely of hot-dung. Before the frame is set on, cover the dung with a foot of bark either new or old. In either of these beds, when the heat is of a proper temperature, the plants in their pots are to be plunged; having the crowns and suckers of the year planted in small pots, one plant in each; and the year old plants in pots of a larger size.

F I N I S.

www.ingramcontent.com/pod-product-compliance
Lightning Source LLC
Chambersburg PA
CBHW021733220426
43662CB00008B/824